FAILURE LEAVES CLUES

the mindset to
turn them into success

Brian Gahan

GLASSSPIDERPUBLISHING

Edited by Murray Reiss
www.murrayreiss.com
Cover design by Rachael Muir
www.blever.co
Author photo by Dale Tidy
www.daletidy.com
Published by Glass Spider Publishing
www.glassspiderpublishing.com

For the three S's and everyone who has given me the opportunity to create with and learn from.

"Saying someone is lost is as crazy as saying there is somewhere they are supposed to be."

–Brian Gahan

Table of Contents

Introduction

I want to explain a very simple technique that could fundamentally change everything in your life. For starters, I'm going to keep it focused on the work environment because that's where this technique was born, but later on, we'll explore other areas of life where the same technique can work for you personally—for your kids, your family, and your community.

I've been extremely fortunate because my career has taken me through a number of industries, always, for some reason unknown to me, in some form of management. But not always at the top. Therefore, I've been blessed to have gotten a point of view where I could learn from my own mistakes as well as from the mistakes of others above me.

I'm a people person, not in the "Hello, my name is . . ." networking groove way but from a fascination with how people work, operate, and manage their tasks and life challenges. I love observing what works and doesn't work for others. I love questioning and exploring all aspects of being human.

I've spent my entire working life contorting myself to be

successful, to turn a business or myself into successes. I spent the last 20-plus years as a partner in a creative advertising agency where I focused on growing and building our business while working with clients to grow and build their own. Over the years, we had multitudes of clients in almost every industry or service on the planet. And the one thing I learned was that all businesses are pretty much the same. They have goals, they struggle to accomplish them, they make excuses when they fail, and they pound their chests when they succeed.

Advertising—for you who don't know—is the business of helping others achieve their business objectives by motivating and/or changing human behaviour. In short, it's all about goal achievement. Our business had its own goals of growing and making lots of money. Our clients had their goals of growing and making lots of money too.

All sorts of sayings were thrown around to express these goals: *Move the needle. Put bums in seats. Grab those eyeballs. Get exponential growth!* And over the years, I did experience some success, some growth. But if I have to be really honest, what I mostly experienced was failure. Lots and lots of failure. Mine, ours, and others. And over a long enough period of time, this failure became repetitive, habitual, over and over again. And terms got thrown around like *fail faster, cut your losses, learn from it and move on, you're going to fail a ton so just do it as fast as you can.* But what I started to uncover was a systemic lack of desire to learn what made this repetitive failure keep happening. Instead, I saw the rise of the mantra: *Don't do that again. Next!*

Fight or Fix

This all came to a head for me a few years back while sitting in a strategy session. The session had to do with a senior player experiencing a state of fight, flight, or freeze when faced with evolving changes that needed to happen and behaviours that needed to change.

I said, "What about fight or flight or fix it?" but the highly paid professional running the session said they didn't like the word "fix"—it didn't feel good to them. When asked, the others in the group agreed, abruptly ending the discussion.

I was silenced but also perplexed by this reaction and the example of groupthink. What is wrong with fixing things, I thought. Or is it the word *fix* that's the problem? We like words like *overhaul, revamp, upgrade, improve,* or the new and ever-so-trendy *pivot.* Is that because they seem like fighting words?

But are they?

Let's do a little case study. We'll call it "fight or flight or freeze." We are all familiar with the saber-toothed tiger and the cave person. We all know the supposed outcome of the fight and freeze scenarios. Either the tiger dies, or you die. End of story.

But for example purposes, the flight scenario is of interest. If the cave person flees, temporary success. But if the tiger tracks them back down, they're faced with the same scenario all over again. What if the cave person analyzes the failure to permanently escape? What if they ask, "How did the tiger find me again?" And if they learn to cover their tracks and hide their scent well, they may actually accomplish long-term lasting success. Hence fixing the recurring failure for good.

You see, what I came to understand is that *failure leaves clues*, and if no one picks up on these clues, then failure leaves seeds for more failure. If we believe success is just stopping failure, then focusing on these clues would be the fastest and most effective path to success.

This brings me back to the simple technique I referenced off the top: It's called 90-90-ONE.

90-90-ONE is a very simple, easy to learn and use technique and mindset. It has nothing to do with success, as such, but it will ultimately lead you to it. It's a process that involves getting all the inhibitors to success out of the way. It's about identifying, focusing on, and improving what you have control over. It will reprogram your brain and change your life!

Chapter One: Goals and the Excuse Economy

Are We SMART Yet?

I've come to realize that success has an Achilles' heel, and it's called "our goals." If we dare to look into the shadow of our goals—into their failure—we will find there the true key and building blocks for our future success.

But to fully understand this we have to reverse engineer our idea of goals, how we work with them, what to learn from them, and understand why they aren't working for most of us. How our focus on making them over and over while placing little attention or value on the clues lying in the ashes of our previously failed goals is keeping us stuck in a loop. I call this the failure loop. Let me explain.

Okay, take a deep breath. I promise not to traumatize you too much.

We're going to think back a bit. Everyone has a time in their lives when someone led them through a goal-setting session. Whether it was at work, in school, or from your favourite online performance guru, we all got some tips on accomplishing things, right and fast. Remember that feeling? How did it go? Was it disappointing? Did it suck?

Why? Because most of us, deep down, don't like goal setting. It's not just because the person leading the exercise may have been annoying, with their endless judgmental processes and acronyms. Or that most of the techniques for goal setting are based on language and strategies that haven't been updated since the 1960s. All valid objections, but I propose it's because there is little or no support and guidance on how to deal with what I refer to as the shadow side of goals. On how to work with the failure side of goals. Other than constant encouragement to repeat the same process and set more goals.

So let's look at traditional goal setting. It's kind of crazy.

For starters, your goals must be SMART! Remember that acronym: Specific, Measurable, Attainable, Relevant, and Timebound. Oh, but make sure they inspire you; put them in writing; write an action plan out for them. Yes, really, they say to do that. And my favourite—stick to it! But before you start . . . Be sure to pick a goal that motivates, but never, ever before you have identified your purpose. And here's the real kicker: once you've done all this and had the strength to start down the path of said goal commitment, and hopeful achievement, along come all the reasons why they FAIL.

Believe it or not, there are more reasons for your goals to fail than there are ways to ensure success. Figures, right? And mostly they are all gut-punches below the shame belt. Let's take a look at the reasons you failed:

1. Fear of success and or failure
{So basically, you're scared.}

2. Lack of understanding about the goal-setting process
{Please refer to the above. It's really not that complicated at all, just get your act together.}

3. Lack of commitment to the goal, and its fun cousin — inactivity
{What, did the process of starting burn you out?}

4. Analysis paralysis
{But hey, can't be the personality types that like acronyms getting caught up in their love drug, could it?}

5. Lack of a real destination
{Yes, all goals must be based on reality. A goal to do something

crazy, that's just crazy.}

6. Failing to plan
{Okay, I get it, back to that Action Plan thing.}

7. Having too many goals
{You greedy, greedy little overachiever. Shame on you.}

8. Feeling unworthy of the end result
{Again, more shame—just keep heaping it on.}

9. Garbage in, garbage out
{Okay, what if I told you it's pattern in, pattern out? But I'm getting ahead of myself here.}

10. Your lack of motivation to change
{Last, but never least, the huge shaming, blaming finger. Are we done failing yet? Not feeling so good? It's because you're not doing them right! They have to be SMART! You didn't write them down or burn them in a ceremonial bowl in the backyard, whatever. Bottom line you failed. No goal for you!}

It also sucked because goal setting is usually the start of some form of restriction or forced behavioural change. You're stopping something or starting something that isn't easy and doesn't come naturally. Like, I will lose ten pounds, or make 20 cold calls a day. Yuck! And because deep down we know there are a ton of reasons for why we will most likely fail or, worse, give up, we are doomed before we even start.

Again, don't believe me, just Google it, or think back to your many New Year's resolutions. There are just as many books, blogs, vblogs, and articles about why you failed to make them work as there are on how to do them right. Yep, and they all point to the fact that it's because *you* did them *wrong*!

Now don't get me wrong, I truly understand and believe in the importance of having desired, intentional outcomes, like losing ten pounds or running a marathon, but when it comes to success, goals are not the be-all and end-all that they have been masquerading as. They are not the magic portals to success. There is something else going on, and that something else is often ignored and hidden in the shadow side of the goals we all set and reset time and time again.

The pattern of goal setting and the 90-90-ONE tool—but even more importantly the mindset I will teach you—help you focus on and mine the shadow side of your goals for true accomplishment and success. But like most shadow stuff in life, we avoid looking for it, acknowledging it, or even talking about it. Until now!

The Work on Goals

There's a ton of amazing work being done on why goals alone don't work for success. And let me be clear, I'm not a professional researcher. I'm just someone whose experience with failure (in many careers and schooling) resonates with the works of people like James Clear, who explains in his book *Atomic Habits* that goals don't work if you don't focus on the system of achieving them. He points out that the goal of winning a sporting match is the same at the start of the game for both the eventual winner and loser. So having the goal means nothing, the *system* of how you execute accomplishing that goal is where the gold is.

And that's exactly what my real-life experiences have taught me. Every businessperson has the goal of being successful, every salesperson has the goal of making more sales. But does focusing on goals alone and the process of making them, make them SMART? Does being held accountable? Or celebrating your goals? Reviewing them? Posting them up or writing them down? These are all good, but if you're not looking at the system or pattern of how you manifest your goals, you will get caught in the infinite loop of making more and more of them to cover up that huge hole that is building in your confidence—that feeling that you keep failing.

Time out! To give us all a little reprieve from the stress and magnitude of this goal thing, I love to reflect on this quote from Marie Forleo: "Most great things that come into your life did not exist when you were goal setting."

Right? Doesn't that feel better? Maybe these goal things are a bit overrated when it comes to real, positive growth and what really matters in life, but even so, there can't be any downside to having lots of goals. Or can there?

The Rise of the Excuse Economy

So what happens as we just keep making goals upon failed goals while never looking into their shadow side to learn what is really going on, what really needs to be addressed? We enter what I like to call the excuse economy. We all trade in it. Some traders are better than others, but a goal-driven organization that does not focus on the systems at both the creation and failure side of goals ends up fueling and building their very own excuse economy. You know what I mean?

We've all been exposed to it at some point. A goal is not being reached, and over time the participants learn what excuses are more valuable than others to justify its demise. John in accounting knows he can pull out "Year-end has been a real mess, but hey, money and numbers are way more valuable to the organization than me achieving this goal, right?" Or "Of course I'm focusing on my goal of running a marathon, but the kids come first, and they've been going through some tough days at school." And then to make matters even crazier, the excuse economy, left unchecked, is single-handedly responsible for the rise of the corporate buzzword "accountability."

"How do we make people more accountable?"

"There needs to be more accountability here!"

"How can my light saber of accountability break through your excuse wall?"

I don't know about you, but I've spent endless hours I'll never get back sitting through discussion upon discussion on the need for more accountability in the workplace. But like most trendy corporate-speak, very few people can even define the term. When pressed, it usually boils down to getting people to do what I want, when I want. So, how do leaders get more of that? I know, let's set up the goal of being more accountable!

Does the infinite failure loop of goal setting, excuse economy, and cries for accountability sound familiar to you? Well, 90-90-ONE is going to break through all of that. You can count on me!

So, for all intent and purposes can we agree that goals don't work, or at least not as well as we have been told they do? Not because the concept of goal setting is wrong, but because the structure of goal setting is problematic, and could be covering up fundamental flaws that will eventually undermine a goal fulfilling itself and building success. Are you at least open to that premise? If you've gotten this far—and your picking up a book about failure tells me that you're not a goal-crushing machine—your mindset might be ready for a shift.

But before we dive into 90-90-ONE and how it works, let's look at another trend in the human desire to fast track over failure to success. And that's the multimillion-dollar concept of mimicking it.

Chapter Two: Success? Mimic It!

What's In Their Coffee?!

Okay, you want success. You want that glorious feeling of achievement. You're tired of being in the dark, bumping into failure, and aimlessly grasping for something that looks like and feels like success, but in the long run isn't. Your goals aren't working and time's running out, so you think *screw it*.

If you haven't achieved success, it's time to mimic it!

You see, most successful people—and by successful I'm using the North American ideal of rich and famous—do things that seem impossible to the average person. Therefore, we are left thinking they must know something or have a trick we don't yet understand and if we mimic what they are doing, if we do something like them, then maybe, just maybe, we'll find success like theirs. Success must leave clues, right? That's what we're sold.

So we take a look. What do the people that have it do? Must be their habits. Something simple that we can adopt. How long do they sleep? And when? Do they meditate? How?

What do they eat? Read? What? Are there lists, mantras, affirmations? There must be a course! The options are endless. It's overwhelming and it's all quite maddening.

Their success formulas are a multimillion-dollar industry trying to convince us that if we act like them, behave like them, think like them, hell, become them, then we will have what they have. Just look at some of these titles from a quick Google search. Do any sound familiar? *15 Secrets Successful People Know About Time Management; 21 Success Secrets of Self-Made Millionaires; Millionaire Mindset; Habits of the Super Rich; Millionaire Success Habits; High-Performance Habits; Millionaire Habits; 7 Habits of Highly Effective People.* Oh, I get it, it's secrets. No, it's time management. No, it's got to be habits! The list goes on and on, and most are compiled and served up by complete strangers, never the owner of such life-changing secrets and habits.

One of my all-time favourites that is often served up by the millionaire themselves is the trend in performance/endurance sports of what goes into their morning coffee. And surprise, surprise, it isn't coffee! You see, you've been drinking coffee old school. That's why you never lost the pounds, built the muscle, or couldn't finish the ultra-marathon. Where's the lard, butter, bacon fat, bone broth, matcha, maca, chili, or whatever the latest trend is? See, that's why you're failing! Come on, be like me, you have a goal, don't you? Get serious with your morning cup of joe! I like to call these performance nutrition trends "Popeye Trends," but that obviously ages me poorly.

Whether it's the morning coffee or some new 5 a.m. ritual, in the end, when I think about all the pressure to mimic, I like to turn to bestselling author and internationally renowned speaker in the fields of human consciousness, spirituality, and health, Caroline Myss, for something a bit more soulful and way less anxiety-inducing. She sums it up simply by saying that if it's not *your* path, trying to follow it creates an inauthentic life accompanied by all the stresses and anxiety that go with that. Then add to that the potential downside of misguided projection that Shawn Corey Carter, better known as Jay-Z, speaks about from his personal experience. "Don't listen to anyone. Everyone tells you how things worked out, but it worked out for them that way. So don't listen to anyone because their experience is unique to who they are. I walk in every room as myself . . . proud."

This doesn't mean we can't try new stuff or add things to our coffee but mimicking is not the missing answer. We have to do the personal work of looking at our unique selves, what we are doing, and work with what we've got.

Why? Because mimicking is a "should" mindset—"I should be doing something differently"—and shoulds are pretty much at the bottom of almost all shame and blame. Not a good foundation for healthy growth. And let's be honest, is success not just a really bad word for growth? And by growth, I mean personal growth—the successful kind!

Getting Personal

Now, don't get me wrong, I believe in good habits. We can all agree that if someone goes to bed drunk every night, fails to get up to the alarm, and is constantly losing important jobs because of it, then changing the habit of drinking all night will most likely lead to success for that particular individual. But, once they become a millionaire, concocting a book about their habit of getting up to an alarm clock is not going to be that helpful to others unless they too have the exact same personal issue.

This is why I love 90-90-ONE. As you'll learn, it goes directly to and stays focused on the personal. You can't fix what you don't have, and learning what Richard Branson does to fix his issues or the latest tech billionaire's obsession with intermittent fasting may be entertaining, engaging, or just plain voyeuristic, but in the end, it won't help you one bit because you're not them. You have your own authentic issues, your own relationship with failure, and without a personal approach, all the mimicking in the world won't get you any closer to your goal or success. And by that, I mean that every desired outcome is personal to the individual, and like we talked about in the last chapter on goals, there are very personal things that get in the way of you and I achieving our desired outcomes.

This is *not* about not having a mentor, trainer, nutritionist, coach, or someone that helps guide you through life both personally and professionally. Whatever gets you up, inspired and moving is personal and none of my business. I'm not here to tell you what you should be doing; I'm offering up a process and mindset of how to accomplish preferred outcomes by looking through the lens of failure. And as we will explore in more detail, failure, especially in North American society, is something that is skipped over and never acknowledged until there is finally success of some sort. If you don't believe me, just think back to all the post-mortems you've been involved in. How did they go? Was there an embracing of anything that went wrong, or just a desperate need to move forward?

That failure thing—we avoid it like the plague. And no one more so than someone with their eyes fixed on success.

If you're onto your fifth book like those mentioned above or constantly watching YouTube for the next best tip or habit to mimic to trick success into happening, then it might just be the perfect time for you to take a deep breath and decide to focus on your own growth and the accompanying clues. It's time to embrace this thing, this so-called failure. Gulp!

Chapter Three: A Bit About Me

Failing Into Writing a Book

Who am I to write a book? Especially one about success when most of my professional life I have navigated through failures of one kind or another. Like most of us have. Many of which can be spun into examples of hard but necessary work, you know, "the school of hard knocks" kind, or "the grit that makes the foundation for eventual success" kind — but still failures in the end.

So why me and why a book about failure? Well, for you to truly understand how I ended up here writing this book, I need to go back and explain a bit. You see, I've always been someone who believes in growth. I have instinctively known that "better" is out there, in our lives, our careers, and in the world.

Everything can evolve into something better for everyone. It's a feeling I have always had and that has constantly driven me. It has fueled my multiple career path changes. It's given me the courage to leave behind situations that others couldn't believe I could walk away from. It has given me the understanding that success does not come from just saying it or claiming it but from constantly refining and evolving the journey we are on. It's about constant learning and growth.

Taking some leaps of faith and following my curiosity and heart, I found myself at 37 years old starting in the advertising business. As I previously mentioned, this is a business of goals, but to truly understand it, you have to know it is fueled by massive ego. But then again, all business is in some way fueled by ego because it's all about competition, winning, and survival, which are all in the realm of the ego mindset. And the one thing the ego truly hates is failure. It hates it so much so that it denies it and remakes every failure out to be some form or stage of success—winning! And if you are winning why would you ever need to look at failure? Failure is just a hurdle that needs to be overcome with willpower. Terms like "step on the gas," "pedal to the metal," or "we need to be *more* productive," are all used to deny any issues and just push through with more of what we think is working or will eventually work. That's ego at its finest, in the world of most businesses.

So, you take my personality type that's always asking "How can we improve this? Is there a better way?" and drop it into an environment of massive ego with its constant boasting and need to win at all cost, and what comes out the other end is a person—me—who has witnessed a ton of failure swept under the rug of mediocre to moderate success and been left with an insatiable fascination with why we choose to try and grow without ever looking at, or acknowledging, the downside of those efforts.

Now, I don't think I'm the only one with this personality trait and in no way am I trying to say that I am similar to some of the great business success stories but I do believe that this trait is common to many of them and has driven the likes of Steve Jobs and others who constantly seek to improve upon . . . well, fill in the blank. So with that aside, I'll give you a case study.

In my time in the agency business, we had lots of "successful" clients, some because of their smarts and hard work and others because of their timing but I was always fascinated by how they were spoken about. They were successful—made money—and therefore they were doing things right, perfect, full stop!

I can remember one time engaging a colleague in a conversation around this very idea. I said what if this particular client just got the timing right while doing a ton of other things horribly wrong and because of this timing they were making buckets of money. But if they fixed some of the things they are doing wrong, could they not be making even more money? Hence, are they failing at being *as* successful as they could be, should be, by not addressing all these issues? Does being so profitable mean there are no areas where they could improve and grow? To which my colleague responded, "Come on Brian, they're doing things right, they make a truckload of money." Case closed.

I believe we are starting to see a lot of this unfold around us. The ways institutions and businesses operate and behave are all coming under scrutiny and they are being asked to fix the broken bits or perish. Past and present success will not give them a free pass on the need to be better and this is what 90-90-ONE promotes—the opportunity to fix and improve the broken bits before they get the chance to reseed and become even bigger, repetitive failures. But in the world of the ego, another word that is hated is "fix" because fix means you have to take responsibility and change. And responsibility is the ego's kryptonite. But more on that later.

Okay, back to the bit about me. Over the years I witnessed time and time again how companies, individuals, and teams would jump from one attempt at success to another rarely looking for answers to the crucial question: Why did we fail? Instead, they'd scorch the past with a flame thrower to ensure no evidence survived, often scapegoating and eventually dismissing valuable team members in order to hide any evidence that, hey, maybe we messed up and need to fix something here. Nope, let's just put the foot on the gas and drive as fast as we can into the next project, campaign, effort—but, more than likely, the new next failure.

I've also been exposed to my share of leaders that refuse to look at or perform reviews on past failures. For example, I was once in a senior meeting talking with a large group of people about the need to do postmortems on why they were losing new business pitches, and the leader, who was a systemic "next" person, abruptly got up and left the room minutes into the conversation. This leader was incapable of even starting a discussion about looking backward. In their mind, it was a waste of everyone's time because "You never look in the rearview mirror. That's for losers and fools. More gas now!" is the familiar mantra for these types.

Does Any Of This Resonate With You?

Are you a middle manager stuck in an organization like this? Feeling the wearying and soul-tearing effect of habitual Groundhog Days of "No one ever learns from anything around here," or "Why don't we acknowledge stuff that's right before our very noses?" Well, that was me and that was where 90-90-ONE was born. Born in a place of ego and chest-pounding, denial, and gaslighting. In a culture of never, ever look back, just keep pedaling or swimming, or whatever you need to do. It became a lifeline for a person, me, who believed we could improve, knew it was imperative for our very survival, and was desperate to have a tool to broach the conversation of, "Hey, that looks messed up, maybe we should fix it before it screws up something really bad." So, at first 90-90-ONE was my own survival technique, it was personal and lived in my head. In fact, it just dropped into my head one day as I ran my morning 10K route into work.

A Creative Side Story

I learned way back in the early stages of my creative career a little survival trick. When you are the creative strategist, your job is to fix someone's problem. It seems simple on the surface. Come up with the idea or manage the team that will work on fixing said problem.

Some call it starting with a blank page, a place many are afraid to go to or be left with, but it can be more complicated than that, and I used to liken it to being thrust into a straitjacket you're forced to wear. It's a sometimes paralyzing responsibility—to have other people's problems literally landing on you and sticking to you as you navigate through your day just because you and your team are the ideas people. Your job being to fill in the blank page so others can feel safe. Oh, and hurry up while you do it.

So, like many creatives, I would be left alone late at night in the studio, long after everyone else had gone home, desperately trying to come up with *the* creative solution for our client's next, new problem. You know, that bums-in-the-seat, move-the-needle, change-this-or-that behaviour thing everyone is so desperate to grasp.

In the early days of my career, I would spend countless hours twisting myself into the fetal position without a clue of what that idea could be. Eventually exhausted, drained, and defeated, I would give up, lock up and head home. My mode of transportation in those days was by bike, the pedal kind. And time and time again, no matter how complex or big the problem to be solved, a creative solution would always drop into my head as I pedaled dejectedly along the late-night, darkened city streets on my 30-minute ride home.

The more I tested this process the more it proved to work. The late, tormented nights at the studio started to get cut shorter and shorter as I learned to trust that all I needed to do was have the problem in my head, along with all the raw data, and then get on my bike and ride home. I replaced all the wasted late-night hours in the studio with a 30-minute ride and a few moments of jotting down my newfound ideas once arriving home. Now, in time for dinner and maintaining a healthy relationship.

So this technique or trick of sticking a problem into my head and then getting outside and exercising worked for me again and again, and over the years it grew into trusting that it would also work even if I waited through the night and used the early-morning hours as I ran into the office once the bike riding turned into morning running to commute.

Now fast forward to this fatal day some 20 years into my career. I'm a partner in a company, managing multiple divisions and their department heads, not only still responsible for coming up with most creative strategies but also operations, parts of HR, and some new business that have all crept under my fold. So what I'm getting at is, lots of problems, issues, or concerns that have to be solved.

Okay, back to the above-mentioned situation of a culture that's built on refusing to acknowledge flaws, work on issues or fix anything that has been deemed yesterday's issue or no longer of use. I'm exhausted and tired of people around me being blind to the obvious. Struggling with how to introduce change, positive change, to a culture of habitual failure and Groundhog Day behaviours. Desperately looking for the Trojan Horse to penetrate this fixed ego mindset. When, on a fateful morning run, the solution dropped into my head. It was simple, it was a technique, it was called 90-90-ONE and I knew I could sell it through and it would work, instantly! But before I got even a few strides further, the realization that exploded in my mind and sent chills through the early-morning air was that this technique was not just a management tool to help me fix some pressing issues, it was a mindset! This technique was a mindset for success because what I had observed and was trying to articulate all these years was the one thing that was being ignored by everyone. The simple truth that *failure leaves clues*.

Chapter Four: Looking into Failure

Got Your Attention?

Yes, I believe failure leaves clues, but before we dive into that, let's take a second to look at this word and the concept of failure.

Some words have gotten bad reputations over their time with us, becoming emotionally charged and taking on other meanings altogether, often to the point of us shutting down anytime we even hear them.

Take the word "ignorant," for example. To say someone is ignorant of something is supposed to mean that they know nothing about it, they are yet to be informed and therefore are ignorant to how something works or what something means. But, in the present day, to say someone is ignorant has morphed to be a reflection of how intelligent they are.

It never leads to a productive conversation, and it never ends well. Therefore, we use language like "they are uninformed," shying away from using the word ignorant in its true dictionary meaning. Failure has suffered the same misfortune. And in the business world, it has become stoked with fear and stigma as I highlighted in previous chapters and we will explore more below.

Let's explore a couple of thoughts around failure. For starters, is success just the ending of failure? I remember the first time this concept was presented to me. Wow, did the lights go on. We are always failing until we have stopped and now we enjoy success.

Success is not the complete absence of failure, failure is more like the straight line of progress and growth where we have failure after failure after failure until it stops, and then by definition, we have success. Therefore, could failure be the verb to the noun success? Are they actually the same thing? Connected linearly like life and death, one begetting the other? By running I have become a runner. By being in, learning from, and working with failure, I become a success.

If there is truth in that, and I believe there is, then all the focus on getting the failing part out of the way as fast as possible seems counterintuitive to the process of creating meaningful, lasting success.

Remember that infamous entrepreneurs' rally cry, "Fail faster!"? Well, they're definitely not focusing on the journey or the process and the valuable lessons they can craft, potentially producing accomplishments deeper and more meaningful than just getting to the end result of a predetermined "success" as fast as possible.

There's A Word For That

We have all seen those examples in textbooks or on the internet of how language can lead to awareness, right? You know the ones where they show you nine squares of green and ask you which one is a different colour and to us, they all look alike. They are all the same colour green, but to the Himba people of Namibia, who have more words for the colour and shades of green than we do, the one square jumps right out at them as being different from the others. But, if you substitute one of the green squares for a blue one, we see it right away but the Himba can't because they do not have an independent word for the colour blue. Another example can be expressed through descriptive feelings. For example, Spanish has more words for and ways to say "I love you" than Japanese, so the Spanish culture is deemed more emotional, intimate, and romantic.

Take this understanding of language leading awareness and turn it on to the word failure. If we have not fully explored failure and how it is part of the growth process leading to success and therefore do not have the language for this complexity, then no wonder we do not fully appreciate and embrace it. With a simple mindset shift of what failure is and its value, we can open up to complexities that are not so black and white when it comes to our own and our projects' growth.

Now, let's take a moment and look at the word success. It's one of those statements that we don't get to use about ourselves. Let me explain: there are a few things that people are not allowed to say about themselves. For example, I'm a good listener, or I'm funny, or I'm a nice person and my all-time favourites, I'm a good kisser or lover. You see, all those statements can only truly be said by someone else. You can think and believe you are funny, a great lover, etc. but it is always up to others to confirm the validity of such skills and traits. This isn't rocket science and most people can grasp this idea but in the business world, full of its narcissists, this concept can get blurry. I'll never forget sharing "you can't call yourself that" with a CEO one day; the look of bafflement in their eyes was so telling. "What do you mean I can't just say something and make it true?" But I'll leave that to another book. For now, what is important is that success is something that is defined outside us by society and culture. And the goalposts can be, and always are, moved. Right now, in our business culture success is predominately based on money and domination. And because you cannot control the goalposts, is it millions or billions that is required? Or with the new consciousness will it shift to the triple bottom line (people, planet, and profit) right when you think you are on track to making enough dollars? Whatever the definition, you are always going to be chasing something that is defined by forces outside yourself that you cannot control. Hence, you're always living in the failure phase of the failure-success polarity. Yikes! And if you're fast-tracking failure, the phase you are always actually in, that has to be pretty anxiety-provoking and stressful no matter how far along the line you define yourself. Now don't you wish we had more words for failure?

Zen on Failure

So if failure and success are part of the same journey and success is outside of our control, then shouldn't we be putting all of our effort into removing our ignorance about failure? Should we not be embracing failure? How does one work with failure, what are its lessons? Can we truly engage failure and understand its complex, scary, dark, shadow bits? Can we fall in love with failure as much as we believe we will love the future success? Not to get too Zen, but how can you love success if you do not love failure if, in fact, they are the same journey along your chosen path?

Here's an example I love of how failing is so different from experiencing failure. Adam Savage, in his podcast interview with Tim Ferriss, said failing is getting drunk and not showing up at your kid's birthday party. That's failing. Yes, another drinking example but they really do help get to the point of things. Adam goes on to emphasize that trying to do something, create something, and having not yet accomplished it is *not* a state of failure—it is a process of creative iteration. If you are going to accomplish anything, before achieving it you are in failure *but* you are not failing. We would never say to someone in medical school that they are failing at being a doctor. But for some reason, we have stigmatized the words failing and failure. Bundle them together and you keep failing until you are a total and complete failure. We refuse to value and analyze this fundamental component of accomplishment. In fact, Adam goes on to say that true creative iteration is the process of following something where it wants to go and not forcing our will on getting to some predetermined idea of the end point—whatever we think success will be.

If we are not comfortable being in failure—working with it, immersing ourselves in it, and seeing it as something to move on from as fast as possible without investigating and analyzing—we are actually "failing" at doing creative iteration. We are failing at bringing to life what we are truly capable of creating and settling for some predetermined idea of what we thought success should be before we even started on the journey. How crazy is that?

Why We Avoid Failure

There is a technique I was exposed to while studying improv theatre that is called "yes and." It's used to help performance storytelling unfold in a way that is not predetermined. The premise is that a group member will always say yes to whatever another group member proposes and then add to it in order to build a scene that is not preplanned or controlled and therefore has the highest chance of being original, creative, and interesting. This same technique has found its way into the business world, specifically when it comes to brainstorming sessions. By agreeing to say "yes and," ideas are built upon rather than snubbed at infancy.

For this next section on failure, I'm going to ask you to embrace "yes and" as opposed to going down the rabbit hole of a predetermined understanding of why things are. So, with that in mind, let's talk about why smart people avoid failure.

The "yes and" piece of that? It's that we need to agree that most people that run companies or lead groups are *smart* people. I know this is hard for many of you to agree with. Especially if you're stuck in the middle of an organization or group where your only survival mechanism for personal sanity is to believe that in fact the leaders are just batshit crazy and when it comes to smart, the ship has sailed long ago. But trust me, once we go over this next bit about failure, you'll start to grow compassion for the state of how things came to be that way.

Let's dive into a bunch of ideas around failure and see why some smart people might avoid embracing it. First is the concept of *contradictory evidence*. Studies show that once we believe something, we develop a bias against anything that contradicts this belief. In David Epstein's book *Range*, he references a study that outlines this concept when it comes to politics. Once you believe something or someone is a certain way, good or bad, you will automatically reject any evidence to the contrary. This is so evident when you think about some of the powerful scandals that have happened over time. Whether it be the Southern preacher who got caught in a sex scandal or the celebrity found out to have been abusive. The followers or fans automatically reject the evidence that there could be anything wrong.

Now, if success cannot really be defined and the goalposts are constantly being moved and you are always trying to achieve it, then you have to believe you are in the process of becoming successful. Right? And as mentioned, failure is shunned, shed, and nowhere to be seen or acknowledged in this process, so when it comes to success, you have no choice but to brag, exaggerate, and pretend that you are it. And trust me, the ego mindset won't let you down in this department one bit. It lives for and loves make-believe! Have you noticed over your career that almost every person and business exaggerates their success? "Just fake it until you make it" and there is no amount of contradictory evidence or persuasion that can make you switch something from team failure to team success.

Let's add to this the idea or concept of *the simplest narrative*: as humans, we will at most times default or jump to the simplest narrative around something—conspiracy theorists aside. So if the simplest narrative is that we failed, then what led to that failure, and anything it may contain, is rejected as failure too.

How often have you heard, "We tried that and it failed, so we do not need to look into that." Or "there is no value there." Or, even worse, "There is no success there."

So back to the smart people whose only crime is believing that failure and success are not part of the same growth process. Or put more simply: you try something, it didn't succeed, therefore you quickly move it over to the failure column. The simplest narrative is that if it did not succeed, then it is a failure. Now contradictory evidence bias kicks in and absolutely nothing can convince them that failure is full of complexity and that in that complexity may lie the keys to success. Nope, just disregard it and move on. Especially as their ego requires them to shed anything that looks like failure in order to be perceived as successful. And what does any so-called strong leader know to do? You got it— you move on to win another day. Never look back because another contradictory bias kicks in that tells them that this is what a great leader must do. Keep moving forward. It is only through repeated patterns of this that things and themselves as leaders start to look batshit crazy. In the end, that is why some of the most successful people seem a bit crazy because the process of achieving success without ever acknowledging failure along the way seems weird from the outside looking in.

Sarah Lewis referenced this when talking about her book *The Rise* on Brené Brown's podcast *Dare To Lead.* Sarah explains that repeat success can become "dysfunctional persistence." With some moderate success, you can convince yourself that your tactics are always appropriate. And that "repeat success can create an environment that does not support the child-like wonder that leads to innovation." And definitely not the environment to explore and learn from failure and its clues. You see, force of will has run its course and is no longer working so well and as we enter a new consciousness around all this, the business environment is starting to look more and more insane even if we can't quite put our finger on what's wrong. So we default to believing the powers that be are crazy, absolutely not smart, or something even worse.

It's my experience that failure, as the journey to success, is extremely valuable and fertile. It leaves little clues that turn into seeds and grow. But they produce more perceived failure if they're not noticed and dealt with as part of the process of growth and creative iteration on our journeys to success. If we do not have a language for, process of, or even the slightest appetite to work with and be in failure, embracing this part of the journey, then we can never refine the process, pick up on the clues and bring to life the true success that failure is fostering for us, our teams and our companies.

I intend to offer you a new perspective of this metaphysical thing called failure and the clues it leaves. A red pill so to speak. Giving you a technique and a way of working with failure that acknowledges, values, and embraces it. A flowing and working with as opposed to running away from. If you can come along on this journey and shift your mindset around failure you will have a different perspective which opens up new possibilities and changes old habits, which in itself is a much-needed paradigm shift. That shift itself will bring about accomplishments and successes. And, on top of all that you will definitely get a lot more done!

Chapter Five: 90-90-ONE — How It Works

90-90-ONE

The first 90 stands for 90 percent and by 90 percent I mean something that is 90 percent within your capability and responsibility to resolve, change and/or fix. You don't require outside help like the IT department to build you a new web page or run a bunch of cables to deem it accomplished. We all know how those simple IT projects can end up, right? You don't need another person's agreement to allow you to move forward. It's something that you are at least 90 percent capable of accomplishing because it's yours to fix, resolve or make go away. If it involves another person, it most likely will never get done!

The next 90 stands for 90 days. That's your time commitment to getting it done. It doesn't need to take the full 90 days, it's just a deadline that is doable without causing additional stresses.

The ONE stands for the one and only ONE thing that you are going to accomplish, fix, resolve, sort out, whatever you want to call it during the 90 days. Simple, right?

The real power in this technique is the 90 percent. Why? Because we often look at problems around us as being for others to solve, fix or take initiative on and don't see them as something we can be accountable for. Refraining from saying it must be 100 percent opens up one's mind to look at things in a way that creates inquisitive thinking: Where do I fit into this? Is this mostly me? So 90 percent it is! Then, making sure it is just ONE thing, not a series or chain of things, is the other powerful piece and lastly, the 90 days works well in the business world because it fits into that quarterly structure we are all obsessed about. And there you have it, 90-90-ONE, simple at its core but very powerful in its essence.

Don't Believe Me?

So let's go back to the place where I said this was born, the middle-management world of business.

We are constantly being made aware of all the issues that are around us, that we have to navigate through and deal with on an ongoing basis in life, and nowhere is this truer than in the work environment. Just look around: the filing cabinet is not well organized or even in the right spot or not needed at all; the trash can is way too small or far away so people crowd their desk with stuff to then clean off at the end of the day, if at all; and when someone needs to print something they have to get up from their desk just to check on or wake up the printer to go back and press print. The list can go on forever, and these are the things that get in the way of us performing at our best, accomplishing tasks well, and moving towards success. But they are sneaky, they're usually little things that can easily be ignored or brushed off as not being important enough to deal with right now. But our desire to accomplish something is what rekindles them, and with that, they resurface and mess things up big. Time and time again.

I'll give you an example from my days of pitching new business. As the world evolved, and by that I mean the tech world, our pitch decks went from a standard 4x6 format to a "sexier" 16x9 format. And from projecting them onto a screen to showing them on a flat-screen TV or through the internet. But because we were scattered and didn't have a dedicated pitch team, the creation of pitch decks was left to different, random team members, and countless times they would reach for an older, standard formatted deck to pull slides from to build out their new presentation. This often happened because they either didn't know how to look for or didn't have easy access to the new format. And consistently, at the eleventh hour, the entire pitch deck would need to be overhauled to match the updated look. This was never more crushing than when there were multiple presenters, each working in a bubble on their piece of the presentation. They'd come together at the end to build out the final presentation only to learn that there were multiple formatted versions and what was supposed to be a professional presentation looked like Frankenstein himself was going to show up to present. This recurring insanity always came at the expense of finishing and finessing the final presentation content or rehearsing the pitch, and the end result would be sloppy, lost pitches that would eventually lead to, you got it, Groundhog Day. "We'll never do that again" statements were followed by (you guessed it) "Next!" And this is the killer, because there were no postmortems to uncover that this recurring issue was a clue to our failure, nothing was ever done about it and it happened over and over and over again. Like totally crazy, never learn from the past, mind-numbing, over again!

Another example, this one from my restaurant days, reflects how unresolved issues can seriously affect all your business relationships, whether with employees, suppliers, customers, and other partners. Every day the dairy delivery would drop crates of products off at the backdoor to the kitchen, usually between 6 and 7 a.m. The kitchen staff would arrive around 9 a.m. to prep for lunch and bring the dairy products inside and place them in the walk-in cooler. It was how things were done and had been done for years.

During the year, the rising sun would shine down onto the laneway and hit the back door of the kitchen. In the peak of summer, the sun would get so hot and be on the door longer than any other time of the year, giving it enough time to sour most of the dairy products that had been sitting there waiting for the staff to bring them inside. Consequently, the kitchen would be left scrambling during the busy rush, dealing with creams and milks that had, surprisingly, prematurely gone off.

The first approach was to scream and yell at the supplier for sending us a bad product, and demand replacements. When that didn't resolve the issue, we would be forced to change suppliers, accusing the previous one of giving us a bad product and "souring" our relationship.

A new supplier would come on board as fall approached, the sun now cooling and lowering in the morning sky and everything would be great again. Or, at least until the peak of summer arrived the following year, and surprise, surprise, the same issue arose. But, this time we had a new kitchen manager who had worked with this supplier for years at her previous restaurant and had never had a problem with them delivering sour products. Her approach was different. She wanted to look at everything that was happening with the delivery before she would adopt (as we discussed in the last chapter) the simplest narrative—the supplier was delivering a bad product—and commencing the traditional screaming-at-them approach.

She chose to look for clues first. She started with being at the back door when the delivery arrived and while she was waiting she noticed how hot the sun was on her face. In fact, it was so hot she was starting to sweat. She looked up at the sun and right away knew the issue. The hot, intense summer morning sun was souring the dairy products. She assigned a prep cook to come in early to receive the delivery the second it arrived and bring it into the walk-in cooler. We never had the issue again. Our supplier relationship survived to live another day and we avoided the catastrophic issues that were habitually happening from our milk and cream going off.

Some clues are obvious, just sitting there in plain view to be noticed, while others you have to search for. They can have financial, moral, and relationship consequences but they all can be resolved if addressed.

Impressive Numbers!

Now that you get the idea of the types of clues I'm referring to and their potential effect on achieving success, let's take a bigger look at the magnitude of what this simple technique can do for the average business.

To recap, let's start with the individual and their work environment. Look around and make a list of things that are not working for you. The filing cabinet, the book shelf, how you look and sound on the ever-increasing teleconferences, you name it; nothing is too insignificant. Once you have that list, decide what things on it are 90 percent within your control to fix. Or, better yet, spend some time conducting your own postmortem on things that have failed or not gone as well as planned. Look at the role you played in these initiatives, identify what went wrong and ask yourself if there is anything that you can take responsibility for fixing so that it does not happen again. If yes, great! Pick that. But always make sure it is 90 percent yours to resolve. Again, this is the most important part because if the task involves another person or department then, as mentioned, it will never get done. Now that you have picked that ONE thing off your list that is 90 percent resolvable by you, give yourself 90 days to get it done. It's that simple. 90-90-ONE. At the end of the 90 days, pick something else, either from the list or from a new list, and repeat. Good, that's at the individual level.

Now let's take a look at someone who is a department head or leads a team. Take ONE task that your team/department can identify and accomplish as a team, 90 percent on their own, and make it happen over the next 90 days. Simple. The 90 percent is so crucially important here because when it comes to departments, they will almost always pick a task that involves IT, building maintenance, HR, or another department. This 90 percent process that encourages inquisitive thinking is also great for departments and teams because it helps them focus on what is their issues and theirs alone. No finger-pointing, just get it done as a team.

Now let's go macro and look at the big picture significance of this simple tool. For example purposes, let's take a company made up of 100 people. Three of them are in the C-suite: CEO, CFO, COO. Four are department heads. Three are project team leaders and the rest of the team work within that structure.

Now if every group, department, and team were to choose ONE issue that they, as a group, are 90 percent responsible for and capable of fixing—like the above-mentioned presentation pitch deck issue I outlined—and they were to address it inside their own departments, that would break down to one C-suite issue, four department issues, and three project team issues addressed. Then, if everyone in the company took on their own individual issue, like folder structure, garbage can, etc., that would be one high-level corporate issue resolved, seven department/team issues resolved and 100 individual issues resolved. All that in 90 days! Some would be as valuable as why we are losing new business pitches, others could be helping people be a few seconds faster at responding, but all would be something that was getting in the way and hindering success.

This helping people be a little bit faster at stuff is crucially important for all businesses, but it's so often overlooked. Which leads me to a Steve Jobs story. Rumour has it that Steve was frustrated with how long it took the then-new Macintosh computer to boot up, and he wanted it made faster. The development team argued that the work involved was not worth the five seconds they could speed it up by and in fact, no one would even notice a benefit. Sighing that they had more important things to focus on.

Now, *not* one to not get his way, Steve set out to prove that a little thing is a *big* thing when it comes to being successful. So he turned the micro into macro and went huge. What if there are 1,000,000 users of the Macintosh and every day they wait five extra seconds for it to boot up before they got to work. That is over 83 thousand minutes a day which is almost 58 days of lost productivity. His point was that it was worth it, and they did it and, well, we all know that Apple was hugely successful.

So, back to my example of the mid-size company. Do 90-90-ONE every quarter for a full year, and you have 4 C-suite, 16 department level, 12 project team level, and 400 individual issues, potential landmines, obstacles, or failure seeds that the company and its people believe are in the way of success, gone, out of the way, dealt with and not lying waiting to trip you up.

Worth It?

That's 432 things that were there messing around now resolved! That's powerful. And the beauty of the 90-day deadline and 90 percent is that they have a tendency to not be huge time commitments or difficult to accomplish so they're not taking away valuable time from other company metrics. The team is just looking at why things are not working, taking responsibility for what they can fix, and fixing it.

Let me take a moment to share a story that is a great example of the power of the simple ONE. How the search for clues can reveal something that you or your team alone have the power to fix and how your fix can in fact shift things and have a chain reaction kind of impact on the future successes of everything linked to it, thanks to the failure clue that was waiting for someone to acknowledge it. To explain I am going to have to reach back into my days of selling real estate and a time when newspapers were actually something of significance. For my younger readers, a newspaper is something like your phone, but one-dimensional, and what is on it doesn't change, ever. It remains static until the next day when it reinvents itself in an all-new but still static way. How you got your hands on one was complicated and involved multiple people. If you want or need to know more just Google it.

My career journey from the restaurant business to the advertising business involved a couple of stops along the way. The longest one being a seven-year period selling real estate. But not the house on the corner kind. The new, pre-construction kind. Either from a little trailer office in the middle of a farmer's field somewhere out in the country or in a slightly fancier fabricated showroom located in a more urban center selling condominiums and townhomes.

Over the seven years, most of my locations were out in the middle of nowhere. In what was a farmer's field waiting for suburban sprawl to swallow it up. We would be assigned to a specific builder in a specific soon-to-be subdivision and we could only sell to the people that walked through our little office/showroom door. Kind of like a retail store. It was a volume game and my fellow real estate partners and myself got really good at selling the dream. Helping people buy what they could not see, touch, or walk through. We were a special breed, and all we ever needed to be successful was people, volumes and volumes of people. Some weekends we could sell 10-15 houses if we had the volume.

We were good, but the downside was that we were only paid a fraction of the commission that a traditional real estate agent would get for selling a resale property in an established neighbourhood. The reason was we didn't have to do anything but sell and close. No marketing, no recruiting of purchaser or seller, no cold calling—just sell and close. Everything else was the responsibility of the builder's team. All promotion and marketing of the subdivision was their responsibility and they had complete control over budgets and allocation decisions.

Where this became problematic was when it came to advertising, something most businesses wish they never had to pay for and are always looking for ways to reduce the cost of. There's that famous quote by American businessman

John Wanamaker: "Half the money I spend on advertising is wasted; the trouble is, I don't know which half."

These builders monitored their advertising dollars using an industry-standard mini-survey of every visitor coming into the sales office. It was simple, easy to fill in, and had been used for years. The prospective purchasers wanted to get the pricing and floor plans of all the models offered at the location and we would politely ask them to fill in a very quick survey in return. They would always oblige and quickly fill in the answers to get their hands on the price list and models package. The big question asked on the survey was, "How did you learn about our subdivision?" The list would include all of the builder's current advertising efforts. Newspaper ads, radio commercials, signage along the roadway, right down to family and friends. This was before everyone having access to the internet, but even in this modern day of websites, videos, and virtual reality tours, most people will still want to visit the actual location of their potential new home before committing to such an important choice. Giving the builder the opportunity, whether by tablet, kiosk screen, or pen, to learn this valuable marketing data.

This is where the issue lay. Builders were constantly trying to prove that newspapers were not driving the traffic because they were the costliest form of advertising so there were financial motivations to cancel them. But as salespeople, we knew that volume was the sweet spot, and we understood that newspaper ads did work, but the hiccup was that most of these mini-surveys would start with a question like, "How did you learn about our subdivision?" And almost every single person would respond "the signage" because it was in fact the last thing they saw and the thing that told them which country road to turn on or how much further to drive until they got to the location in said field. Anyone that has gone to pick fruit or vegetables at a farm in the country knows these handy A-frame directional signs well.

With this survey data, the builders would eagerly reduce their spending on all other forms of advertising and the first to go would always be the newspaper ads due to their high cost. And once this happened, we would notice our traffic start to reduce and reduce and our sales would start to dry up. To add insult to injury, we would get blamed for the lack of sales because we must not be working as hard or as well because in the builder's mind we still had the valuable signs. What else did we need?

Hopefully, I have painted the picture of our predicament well enough for you to see that we were stuck in a situation heading for failure which we could not control. Not to mention the mind-numbing, endless days of sitting in the middle of nowhere waiting for someone to walk in the door.

We, the sales team, knew the problem—no newspaper ads—and we knew the challenge—garbage in garbage out when it came to the survey data. So we looked for ONE thing that we had control over that we could fix and that ONE thing became these survey forms. We started out by changing the opening question from "How did you learn about our subdivision?" to "How did you end up coming to this area to look for a home?" And the truth flowed forth. Newspaper ads from other builders had enticed them to come into our area and once they visited the advertising builder's site they decided to drive around and see what else was out here and they stumbled across our signs. But it was newspapers that started them on their quest of discovery, and if the builder doing that advertising had employed a great sales team and had good product, we would never have seen them or known they were even close by.

With tons of this newfound data, we were able to convince our builder that they needed to be in the paper to ensure we were getting the prospects to visit our site location first. And with that, sales went back up and the trickle-up effect was huge. Just think about all the people's lives that are made better by living in the new community, the schools that got built, parks and community centers created. All the trades-people that worked on the hundreds, sometimes thousands, of houses. All of which would have had to take a backseat because the initial sales never happened. It's kind of mind-boggling, and as I mentioned off the top, it shows how taking initiative and fixing something that you have 90 percent control over can have a huge impact on your own success and growth but also the success and growth of so many others that are linked to you. Fixing things is a superpower. Becoming aware and having the mindset to notice and take action is within every human's control. We just need a simple technique to support the process.

Okay, so you buy in and you start doing 90-90-ONE what happens next? If you start to clean everything up . . . OMG! then you can reduce the number of 90-90-ONEs. You can do them every first quarter like resolutions, but that might be a bit too, how can I say it, typical and carry some baggage. I'll go into resolutions in a later chapter and why, like goals, they often don't work. But, for now, one reduced schedule that I love is the last quarter.

You reflect on the previous nine months and take an issue that you do not want to start the coming new year with and spend the last quarter making it go away. This is a great year-end team-building exercise as well as giving people something to truly celebrate at the year-end holiday party.

So yes, you can reduce the number of times in a year but never mess with the 90 days, that is what keeps everything manageable. If you go to 180 days or, worse, a year, you will entice people to add bigger and harder to accomplish things to their lists which could fail and ultimately demotivate. If you become lax on the 90 percent you are guaranteed that every individual and department will want someone else to be involved, that's just human nature. Responsibility always feels safer in groups but in groups is where so many initiatives go to die. So no, be stringent on the 90 percent as well.

Even More Benefits

Now comes the fun part—the culture-building benefits of 90-90-ONE. At the end of each quarter, have a pizza and bevi afternoon to celebrate fixing issues that were getting in the way of success and happiness. And I mean really, really, really celebrate the ONEs. Talk them up and make them the big deal they are because, by everyone communicating their 90-90-ONEs each and every quarter, you not only create a culture that takes notice of issues, speaks up, looks around, and makes a difference. Your entire organization's mindset will shift to one of noticing why we are failing, acting upon it, and moving forward towards success. It's truly amazing. Team members will start to learn from each other about common issues or areas they can become observant of, and tackle their own version of, for future ONEs. It's super infectious, in a good way.

I suggest documenting, in journal format, your 90-90-ONEs so that your business has an archive of all the things it has addressed and improved upon over the years. It's also extremely beneficial for new recruits to review previous 90-90-ONEs. It will help them learn what to look for, as well as to understand that this new company takes fixing things seriously and that yes, little things matter. Because you can bet they just came from a previous company where they didn't do just that and as much as some management teams believe that "Rah!, Rah! everything is great here" works, staff are the first to notice what's being swept into the corners or blamed away, and it's empowering for them to know they have landed somewhere that takes care of its own issues, in real-time.

Which leads me to another story from my restaurant days. This one about the value of documenting things. The restaurant business is one of very low margins, high vulnerability, and loads of competition. Some say it's the true business of love over money. There are tons of things that go into a good, successful day or a bad one. And in a seven-day-a-week business, the days can blur into each other and details get lost in translation.

One thing, very old school, we used to do is every evening at the end of the night the closing manager would write a brief outline of what had happened that day in a journal. Starting with the weather conditions, to outside events that occurred to whether there was something on TV that might have kept people at home. Was there proper staffing? Did things flow from the kitchen on time? Were there any complaints, issues, or just plain frustrations of "not this again!"? Was there anything that the opening manager needed to be aware of when they arrived the next morning? That sort of thing. What this written history did, was it gave everyone an instant snap shot into the business, what was affecting it, what might lead to a good day as well as to a bad one. You could see patterns and make real changes to the business.

An example that exemplifies this benefit involves imported beer and accusing innocent people of theft.

One way of controlling theft is to monitor your cost of goods sold. If the percentage suddenly changes, then you could have an issue with a bartender, server, or other staff member stealing from you. In this particular restaurant, this red flag would go up at the end of each summer when it came to beer cost. There was always a slight increase in the percentage cost of beer over the summer months and eyes would be turned on the summer temp staff and the question, were they stealing? Not a ton, maybe a couple of beers at the end of every shift, but enough to trigger this slight percentage increase in cost over the summer months.

That was the ongoing consensus and management scrutinized the team as such. Watching and waiting to catch them in the act. Desperate to prove their hypothesis. But this is where the journaling comes in. Documenting the daily comings and goings and issues that arose, led to noticing there was a problem with staff mixing empty imported beer bottles with the empty domestic beer bottles which were being cased and stacked up by the back door for pickup and a subsequent deposit refund. At that time, imported beer bottles did not get refunds, and were just recycled; they were never to be mixed in with the domestic bottles that were cased and placed outside the back door for the refund. You can imagine the drama, big burly beer delivery folks tossing and discarding all the unwanted imported beer bottles into the back laneway. This often led to broken glass needing to be swept up and drama between the kitchen, serving, and bar staff on who was responsible for taking care of this hazard.

You can imagine the tension, right? And yes, you can imagine if you were the manager that evening you would write it down in the "crap, this is becoming annoying and staff need to be trained not to do this" section of your nightly recap.

So down in the nightly journal it went, not just once but a number of times, which led to the awareness that this problem seemed to increase exponentially in the summer months, which led to the discovery that in the summer we were selling more imported beers. Why? Because people that only drink beer when the weather is hot, tended towards the fancier imports—this was before the local craft beer boom of today. So, because of this trend more imported beer empties were finding their way into the stacks of domestic empties and because of the reduced domestic beer volume we were also not receiving the same dollar volume for refunds which accounted for the increase in overall cost-of-goods-sold percentage for the beer category which, in fact, meant no one was stealing!

So other than being an example of how the restaurant business is all about watching your pennies to stay afloat, this is a perfect example of what a huge potential cultural landmine it would be if management maintained an opinion that the summer temp staff could not be trusted. And how documenting and looking into issues creates a much healthier culture.

Okay, now back to 90-90-ONE and the benefits. With the likes of 432 issues removed, improved, or dealt with, what more could you ask for? Well, let me tell you some of the other benefits of this technique. Yes, it's great at removing obstacles, it fosters accountability—you can count on "me-ability." It will build teamwork and culture. But the biggest surprise side benefit is people development. If you are a leader, team lead, HR manager, or run the company the most valuable thing you can do with your time is learning how your team members tick. And this is a hugely beneficial by-product of 90-90-ONE.

You know that saying, "If you want to learn how someone works, give them a simple task"? Well, how about pick ONE thing that you are 90 percent capable of resolving and do it over 90 days kind of simple task?

Chapter Six: Personality Types

Managing People Is Really Hard!

I touched on this in the Introduction: most of my career, no matter whether as a teenager in the restaurant business or as an adult creative strategist in advertising, I found myself in the role of managing people and the people that managed people. It was never a conscious choice, I just gravitated into these roles. Never with the desire to have power over but more in the vein of taking on responsibility for. Coupled with never wanting to be micromanaged on the "how," this often led to me being put in charge of things, projects, divisions, and companies.

Along with this journey came a ton of management training, hundreds of books read, videos watched and podcasts listened to, and loads of conversations on the newest how-to and the best efficiencies. You name it. I did my best to absorb it and process it through to implementation.

I also had the crazy fortune of getting thrown into the chaos of learning how to manage creative people—which could definitely be the entire content for another book. That experience exposed me to the subtleties of management that do not belong to the more left-brain structure of "this is how it's done!" You know what I mean: the performance review mumbo-jumbo mindset of "just have clear expectations" that is like throwing someone a candy lifesaver when they have fallen overboard. Right name, wrong result. Again, another book.

Now, the one thing I did learn and know for sure is that human beings are extremely complicated, and outside of the military, in a free society, you can't really demand they operate according to your plan. Therefore, I spent my career searching for ways to improve upon the management of people, and what I learned was that the more you know about how someone ticks at a very personal, individual level, how they approach and tackle challenges, the easier it is to be out in front of them and hopefully manage and support them well. Especially for the people that you manage that are managers themselves. Why? Because there's a ripple effect. How and what you do with them will definitely trickle down into how they are with their reports and what they do with them.

I also learned that there are two types of people that take up management roles; there are the ones who were more or less programmed from birth and never really made a conscious choice to be on that team and then there are the ones who did. You know the ones? They see the prestige and crave the power over more than the responsibility for. These are the difficult managers, we've all worked with them, been managed by them, or, unfortunately, had to manage them.

This chapter is not going to resonate well with them at all. Why? Because they lead by design. They live and breathe all the traditional buzzword stuff of the left brain—how-to, with little to no self-awareness or attempts at truly getting under the hood of their team members because it would mean they would have to open up and look under their own.

We've all been exposed to a lead-by-design manager/leader who sends around the latest expert's list of the top eight things to do for . . . insert here any of the following: team building, collaboration, creating an innovative and/or safe environment, being more productive and/or accountable . . . you name it. There is always a couple of things on the expert's list of eight usually number six and possibly seven that everyone knows this leader does or doesn't do but the leader, themself, is blind to it.

Maybe they only read the first couple of points on the list, up to number four or five, and really wanted to show leadership by design and sent it off in a hurry without ever reading to the end or reflecting on the list in any kind of personal, meaningful way or, the first or second thing on the list is something they have been demanding there needs to be more or less of. Either way, their effort usually ends with the opposite of the desired effect. Everyone else on the team or in the company, aware of this leader's blind spot(s), thinks, "Well, that's why we can't have and never will have" ... insert here again because they don't even have a clue that they do or don't do #6 and #7.

Yeah, that leader. Well, they won't connect with the stuff I'm about to share as one of the crucial benefits to 90-90-ONE, in fact they most likely couldn't muster the courage to pick up a book with the word failure in the title, but for born leaders and those aspiring to be better leaders, this benefit of 90-90-ONE will be invaluable in growing and developing your team and might, in fact, help you uncover if you have some of those types of toxic leaders working within your team, structure or company.

This bonus benefit, which will help leaders manage their people and teams, comes with some interpersonal observation work. People are complicated and things never go easily or as planned, but if you embrace the simple 90-90-ONE structure it will automatically bring to the surface lots of fodder on how to work constructively with your people and teams. You just have to be open to noticing it and working with it. Let me explain.

I've broken the people you'll be working with down into six personality types. I use the word "personality" because our personalities can be reduced to how we think, feel, and act—which is a great way to approach managing others. Are there clues to how they are thinking, feeling, and hence acting? Any glimpse into this can be huge so reverse-engineering the 90-90-ONE task into a thinking, feeling, acting personality type is powerful and helpful. With that in mind, meet the Pleaser, the Show-Off, the Complicator, the Avoider, the Simplifier, and the Personalizer.

A quick note: people can take on parts of any of these personalities and float between each depending on the actual ONE they have chosen. Just as our personality can shift if we work on how we are thinking, feeling, and acting. How you work with your team will affect the outcomes of this and future ONEs, and depending on the ONE, people can display different behaviours. What I'm getting at is that in no way by bucketing people into these six personality types am I suggesting this is a fixed process. People are fluid and so should your interpersonal observations of them be. Always be present!

The Pleaser

Characteristics:

This individual wants to get it done, and fast! They want to show that they are on board, up to speed, and efficient. Tasks need to be easy-breezy as well as something that will impress on the surface. Completing the task is more important to the Pleaser than the task itself.

How to Spot One:

Like the Show-Off, outlined below, these individuals will come to your attention. They will eagerly tell you what they have picked and update you on its progress—constantly. Their goal is to please and efficiency is how they do that. They will pick items that are easily solved over real problems that are stopping bigger goals from happening.

Over time, they have a tendency to run out of ONEs quicker than other types because they create the problem with the solution already in mind. "I'll pick this because I know I can get it done fast," kind of problems. They already know how they are going to solve the ONE before they pick it. They have the steps planned out and would never bring forward something that they had not already figured out how to solve.

They never start out by searching for a problem/issue; they always start with seeing a solution. Remember my sour dairy product example in the last chapter? Well, the Pleaser won't uncover those types of deeper issues.

What is Really Going On:

As managers, these types can cause serious morale issues with their teams. They need to look good, and the 90-90-ONE process becomes about them. They pick ONEs that make them look good, not necessarily the rest of the team. They put themselves first, and as a result, their team can feel unsafe. These types are masters at managing up so at first, they can be hard to spot if they have already won you over.

How to Help:

Ask questions about their ONE, help them understand that it is the depth of thinking around the issue that you value. Praise the substance of the ONE, inquire about how the team is participating in the process. How they uncovered it as an issue. Ask for details, the back story. Tell them it's okay to pick something that might eventually take all of or even a bit longer than the 90 days to accomplish and watch their face and reaction.

Not getting it done quickly is their breaking point. As their manager, if you can make it clear that you value the team above all else, the Pleaser, by their very nature, will adjust to the benefit of the whole.

The Show-Off

Characteristics:

This type is the opposite of the Pleaser. To them, the task is the most important and outweighs the ability to get it done. This type is ambitious and their desire to impress everyone comes through from the very first ONE they choose: "Look at me and how I pick such important and well thought out things."

How to Spot Them:

It's easy, they will always pick and present their ONE as something that reflects a bigger business goal of their superior and/or organization. For example, their goal will sound like they care first and foremost for the company's goals and objectives: increasing sales and revenue, cutting costs, improving productivity. Their ONE is an existing goal of the organization but masked to look like an original ONE.

Are you impressed with them yet? Listen to their ONE carefully, does it sound familiar in some way? Like something we are already striving for?

What is Really Going On:

This type is not an individual thinker or problem solver. Their ambition clouds their ability to fix and learn from mistakes. These types will come across as positive, but they fall into the trap outlined in Chapter Four of not learning from failure. They don't look for clues for why the company's goals are not being achieved but, instead, double down on those goals. Likely their team members see them as a bit spineless and self-promoting and it will be hard for this type to build a solid team because they are at the whim of upper management and not in the trenches with their team. They are reactive and not proactive problem finders.

How to Help:

Remind the Show-Off of the importance of being seen as an original thinker both inside their team as well as in the company at large. Impress on them the importance of being someone who is solving problems that everyone below them sees and is concerned about, and not worrying about impressing the powers that be. Allow them to feel safe exposing their concerns and insecurities because they are going to be new to this type of independent thinking. Your effort will be worth it.

Once you turn their thinking around, with their sense of pride still intact, they will champion the ONE and the 90-

90-ONE process, because their ambition is still to show off. But now they are heralding their team's joint effort.

The Complicator

Characteristics:
Always starts out enthusiastic and positive to both concept and approach. Comes across as efficient and wanting to get things done. Eager to finally be given the authority to solve some pressing problems.

This type is similar to the Personalizer, outlined below, in that they take the process very personally. What they choose as a ONE really matters to them. It's important, it's been an issue for a while and now they can finally solve it. It's something that really matters! Because it is the importance of the ONE to the Complicator that triggers the complication.

How to Spot Them:
This type will almost always pick their very first ONE to be something that pushes the envelope. Right out of the gate they will test the limits. If there is a budget, their ONE will be right up against it or challenge it. If they have picked something really personal, there may be some grey area as to how it is really relevant and a real ONE. As their

manager, it may feel like there is something bigger at stake, another agenda, and there usually is.

What is Really Going On:
In the end, it's about sabotaging the process. For example, how can they complicate this so that they can prove that their ideas are not valued, no one listens or takes them seriously, or whatever may be their internal dialogue. By picking something that challenges the budget or others' perspectives they can create resistance and pushback. Now it's a win-win. They can prove no one supports them or takes their team seriously and they don't have to participate in the next 90-90-ONE because the first one never got approved or resolved. The Complicator takes on the 90 percent and then figures out how to make the 10 percent a deal-breaker, whether budget, approach, or philosophy. This type of behaviour can solidify and bond their team members but alienate their team as a whole from the rest of the organization.

How to Help:
With the Complicator you need to be prepared from the start; if you have a budget limit, know you will need to have exceptions. Understand what is really going on and dialogue with this type constantly throughout the process. Help channel their need to complicate into dialogue and

debate, not into sabotage. What this type is truly asking for is this: "Prove that I and my team's needs, concerns, and issues are valued." That is their ONE. That is what needs to be fixed.

Let me break away for a bit and give you a couple of examples, because of all the personality types, this one can be the most destructive to the entire 90-90-ONE process. Why? Because others are watching and learning from them. They are pushing the boundaries and if it's going to break, they will break it. I'll give you a personal example, but first, one about a team leader I'll call Cheryl.

Cheryl is in charge of Research and Development for a midsize company. Her team's job is to make everything better, be it an existing product or service. Her team is also responsible for coming up with the new, innovative ideas for the company's future growth and prosperity. Important team? Yes. Cool position? Yes. But like most growing companies, the business's focus would often be torn away from the new and improved future to present-day survival. During these down periods, there would always be budget cuts in R&D. The company's lack of leadership, foresight and planning would cause Cheryl's team to be limited in its resources as well as temporarily losing its seat at the decision-making table. Everything in her team's world would be put in limbo.

Growing from these periods of uncertainty was a toxic undercurrent in Cheryl's team: No one takes us seriously or values our contribution to the growth and success of the company. All that matters is sales, and only when times are good and the coffers are full do we even get close to what we need to do our jobs properly.

When tasked to come up with her team's first ONE, Cheryl and her team enthusiastically came back with a great idea. Their brainstorming room was in desperate need of one of those smartboards that would record their sketches, creations, notes, and thinking. They would choose it, purchase it, and install it themselves, so it was definitely a 90 percenter, and the only issue was that the smartboard they believed they needed, that reflected their status as innovators and leading-edge creators, was not cheap.

Now, if you have been in the management game for a while, your knee-jerk reaction to "a fancy whiteboard will solve it" is understood. Myself personally, I've approved countless whiteboards over the years that were supposed to solve all the problems but years later hung on the wall still having the original details written on them fading away. But, like I mentioned in the overview of this personality type, the ONE is not the issue, it's "Do you value us?"

With that in mind, and understanding the significance of her ONE, Cheryl's manager agreed to her spending the money and getting the smartboard that reflected that her team was in fact the cool innovation team. It would improve their brainstorming abilities as well as scream, "You are valued!" A real win-win!

End of story? No. Here is where it all gets complicated and why I call this personality type the Complicator—because of what Cheryl does next.

As mentioned, the issue with this personality type is their underlying desire to prove some hidden internal belief. So they will pick something that pushes the envelope. In Cheryl's case, that was the budget envelope. Hoping that it would bring the process to a halt and prove that they were not valued and so would no longer have to participate in 90-90-ONE ever again. But her insightful manager surprised her and gave her the green light. Then, after getting the go-ahead, Cheryl went behind her manager's back to the CFO who subsequently turned the money request down, not understanding that the request was coming from the 90-90-ONE budget.

Cheryl claimed she just wanted to double-check that it was okay to spend the money, all the while making sure that the real story was, "See this isn't working, no one values us, and because of this we are no longer participating in 90-90-ONEs." This narrative is toxic because, as I mentioned earlier, others are watching.

Now, luckily for the company and its culture, Cheryl's manager caught on to what was going on and intervened and got the high-tech smartboard approved after educating the CFO and the other powers that be that it was paramount for the success of the overall 90-90-ONE initiative for this ONE to go through. The teams were shown that they had the power to fix what they felt was broken and that their thinking, efforts, and ideas were valued. This time a true win-win!

So think of the Complicator as a real gift, not a threat. They can be in any department, not just the trendy ones. The Complicator, by trying to break the process to prove their internal belief narrative, can ultimately show others what is possible at the outer limits of the program. As long as you can get out ahead of them and recognize what is happening, you can intervene and turn it into a real positive for the program before it blows everything up. And BTW, 90-90-ONE is not the only place the Complicator is complicating things to advance their narrative, trust me, but for some simple reasons, it brings it to a head right out of the gate. As it did in Cheryl's story.

Now to a personal example about how I complicated my own process of fixing a ONE with—you got it—the solution! Remember back to that bit about me using my running into work as a way to clear my thoughts and come up with ideas, solutions, and insights?

For many years, I would rely on my memory, or I would stop and take out my phone to jot down an idea that came up. And for all of those years, I was desperately searching for some form of device that I could hold in my hand, ready in an instant, that was weather and sweat proof and I could easily dictate my thinking into. I'd learned years earlier about a product that you could dictate into, and it would send you an email of the text transcription, which seemed ideal for my purposes, but it had long been discontinued. Anything else I could find seemed to pale in comparison or was way too complicated, so I continued my search never really finding the perfect solution. As the years went by, the size of my phone grew, as they do, to the point that carrying it in my hand on a long run was also becoming infeasible.

Once I left the advertising agency and started to ponder this book, I realized more and more that I needed something, anything that made it easy to collect my thoughts. I broke down and invested in a very expensive solution—the Apple Watch. It would stay easily on my wrist no matter the weather and I could quickly dictate voice memos of any thought that dropped in while I was out on my runs.

But here is how I complicated things.

You see, my ONE was fixing the issue of not being able to record my thoughts and ideas on this about-to-be book, full stop. The solution was the Apple Watch. But now that I had the solution, I complicated things by using the Apple Watch to listen to podcasts and audio books while running and what transpired was an occupied mind, not allowed to wander, with very few, if any random thoughts of inspiration dropping into my awareness. And even worse, if I did have an epiphany or idea, by the time I paused the podcast app and opened the dictation app, the thought would disappear or morph just like a dream does to the waking mind of morning.

I can't tell you the number of times I would stop to record and by the time I had the watch ready I would end up just recording me stumbling and bumbling on about not remembering what I was thinking of saying in the first place. I'd even start to run again while still recording, hoping that it would get me back into the mind space of the thought, but it would never happen and eventually end with me turning the recording off with a frustrated, "Ah, crap!"

It wasn't until after a lengthy dry spell of no creativity and lost thoughts that I realized the errors of my ways and went back to using the watch for exactly the ONE reason I bought it in the first place. To sit quietly waiting to record the thoughts and ideas as they came to me whenever they were ready to be received.

This is a perfect example of the importance of not only making the ONE a single-minded focused thing but also making sure the solution is not something that can be highjacked away from the very ONE it was meant to resolve.

The Avoider

Characteristics:
This type can be hard to spot because at first, they really like the idea of looking into the problems, using their analytical mind and problem-solving skills to solve some issues. They are so engaged and serious that they want to really think it through and come up with a ONE that is worthy of their time and energy but also will make them look smart. So they go off to think about it. And think about it . . .

How to Spot Them:

Usually introverted, serious, they value thinking every-thing through, thoroughly. They will definitely want time to mull this all over. If there is a regroup, they will have a few things they have been pondering but are still not ready to share or to commit to a ONE just yet. They will flatter the 90-90-ONE process with how important it is to get this right. They just need a bit more time to confirm their chosen ONE.

What is Really Going On:

Two things are happening. One, the person is usually lazy in some way and does not want to spend any energy look-ing backwards. Two, they have strong egos and, as was mentioned, egos hate the thought of failure, so they are try-ing to avoid it—they have enough on their plate and more important things to do than worry about little problems from the past. Unless they can make this into something worthy of their attention, they're going to question the pro-cess and purpose rather than getting down to fixing a trivial problem. Like the Complicator, the Avoider can be a threat to the whole 90-90-ONE process, but for them, it's with their intellectual, superior thinking, and questioning. Others see them as smart, and if they are questioning things, then there must be something to question. Big picture, that is what they do in all the projects they are involved in. They trivial-ize things they deem unnecessary and if they run a department, the toxic trivializing spreads like wildfire. "We are the smart ones, and we don't waste our time." While all along getting little accomplished.

How to Help:

Acknowledge that their thinking is important and very valuable but try to engage other members of their team in the process of choosing their ONE. Don't leave them to themselves to figure it out, remember they are ultimately lazy. Help them see the value in fixing something simple, give them recognition in front of others for their thinking and choices. Their goal is an easy win that looks good to others. Help them by putting value on what they think is just a simple fix. Remember that quote, give the hardest task to a lazy man, they will find a simple way to accomplish it? This is the value of investing in helping the Avoider engage in the process. They can find simple solutions for complicated problems, but you do have to get them started and support them along the way.

The Simplifier

Characteristics:

We all know this type, everything is easy, simple, fast. Right? This type is predominantly responsible for the never look back, never analyze, just keep moving forward culture because it's all so simple to them. Everyone else is the issue, leave it to the Simplifier and stuff gets done. Well, until it doesn't.

How to Spot Them:

"It's simple." Really! They will have those exact words come right out of their mouths from the get-go. Often, they are never really 100 percent engaged, leaving the meeting before everyone has had a chance to vet or share. They get it, it's not complicated, they're on it! But another giveaway is they will definitely pick a ONE that is nowhere near a 90 percenter. Because everything is simple in their minds, they simplify others right out of the process. Until it's too late.

What is Really Going On:

This type is often someone with a very strong ego and a huge dose of ego's sidekick, insecurity. They may be a full-blown narcissist or just someone without a lot of self-awareness but either way, they don't value others and will often just force an end result through sheer will. Often crashing and burning and leaving a lot of dead bodies in their wake. Don't get me wrong, someone that can turn complicated issues and concepts into simple to understand actions is of huge value to any project or organization but, if they simplify steps or, worse, people out of the process it can be demoralizing to other team members and will create passive-aggressive saboteurs to their efforts.

How to Help:

It's tricky because the only way to help the Simplifier is to slow them down, which they will hate and definitely resist. You will need to educate this type over and over again on the importance of the 90 percent and help them understand that they may be tackling something where the outcome is not 90 percent in their hands. You will have to introduce them to the people and steps they are leaving out of their thinking around their selected ONE. Unlike the Complicator, who is ultimately self-sabotaging through not doing something or participating, the Simplifier sabotages something without that being their intent but more because they leave out important players and steps from their thinking. In the Simplifier's case, they end up doing huge damage to the project and the organization by just being careless in how they choose their ONEs. This is a reflection of how they handle most projects and tasks—not so "simply," especially when their ego gets firmly engaged.

The Personalizer

Characteristics:

Independent, hard-working. Not known as nurturers. They get stuff done and don't ask a lot of questions. They take pride in doing what is best . . . but always from their own point of view.

How to Spot Them:

This type will often choose a ONE that seems more like a new initiative and not really a problem fix, even though it will be presented as needed in order to solve a problem. An example would be our filing cabinets are a mess, so we are going to reinvent how we file, not just clean up and organize the cabinets. It's a creative fix, not a practical fix. Another example: the receptionist is not answering the phone politely and is turning off clients and potential customers, so we are only going to communicate through texting and no longer answer the phone. Rather than train the receptionist. The issue is usually presented as an observation of something that is needed. An add-on as opposed to a fix.

What is Really Going On:

This type is common with creative, problem-solving people. Their desire to create trumps fixing. They take an issue and come up with a solution that involves creating something new and—by the nature of how the solution is born—usually very personal to them. Most often these solutions become another problem, but because of their personal attachment to the solution, a wounding happens in the process for this leader. "I saw a problem, I came up with a solution, and no one is embracing my solution," breeds frustration and a reason to give up on more ONEs. "Look at everything I did and still nothing is changing." This can lead to a team that feels like they are always in trouble, being scolded: "We all agreed that we had an issue and I tried to fix it, but no one is embracing my solution." Because of the independent characteristics of this personality type, they tend to take the 90 percent and turn it into 100 percent them, leaving the team's only option for participation being to embrace their solution. Which can lead to a demoralized team and an unembraced solution.

How to Help:
Get in early and help by suggesting that ideas be workshopped with all team members. Try to bring the solution being offered back to the original ONE. Are they truly connected? Work to remove the personal identification with the idea. Help this type understand that it's not all on their shoulders to resolve it alone. Encourage their team to get to the point where they cannot even remember whose idea it was. 90-90-ONE is a great team sport, and for the Personalizer, there is no better lesson to learn from it.

Good Is In the Details

The smallest task can reveal the biggest issue—if we are

aware. That is why I love the 90-90-ONE process. It makes everyone more aware. Really aware. By asking your team to stop, drop and notice, so to speak, you can truly get a glimpse of and become aware of how they think, feel, operate, and act.

"Give a person a simple task and their true nature is revealed."

This is the added overall surprise benefit of the 90-90-ONE tool. A glimpse into the workings of each and every team member, coming to the surface for the astute leader to become aware of, learn about, and use to help develop and guide their team members to success.

Not only do you get a ton of issues, problems and failure seeds resolved you get a window into the complicated human makeup that years of one-on-ones, performance reviews, and motivational speeches will never offer up. How's that for hitting it out of the HR ballpark? You not only fix issues that are tripping the business up and causing more potential failure, but you also fix people issues, too.

Chapter Seven: Let's Get Personal

David vs. Goliath

And by personal, I mean those insane, daunting once-a-year RESOLUTIONS!

90-90-ONE, as mentioned, was born in the heated, stressful world of business and people management, and on many levels was just my own personal survival strategy for getting issues acknowledged, addressed, and dealt with. But through using it, over time I came to realize that 90-90-ONE also provided an amazing mindset for our personal lives. From fitness to parenting—you name the goal. The range it covered was huge. But I'll leave those for another time. Right now, I want to make it micro-personal. So let's look at you as an individual and how 90-90-ONE can help. And by help, I mean of the *David* kind.

Let me explain how you can use 90-90-ONE as your David against the hugely annoying Goliath—that never-ending, soul-destroying, character-assassinating, ever-so-humbling, mother of all goals—the New Year's resolution. Poorly timed, breaking all the cherished acronyms of goal setting, and worst of all, hugely socialized so that everyone knows how you failed at planning, keeping, or even just having one. Entire industries profit from our obsession with having an amazing Goliath of a resolution as well as our inability to keep them. Take the fitness industry or diet programs as examples. They cash in on the masses lining up in January to fail in February, never fully using the facilities joined or products purchased.

How many people do you know who have actually accomplished a lofty New Year's resolution? How many have you accomplished? According to some studies, those that have can be as low as in the single-digit percentages. And that's not taking into account all those people that adjust their resolution midway to fit their lack of progress. You know, those positive attitude folk: Those "I'm really proud of myself for understanding that my goal to lose 20 pounds was just not right for me, so I'm good with the few I lost" folk. And then there are the people who are going to lie when it comes to these types of survey questions. So, with all things considered, it's pretty discouraging.

You see, the issue is not with the desire for better. The issue is with the timing and magnitude of these mammoth once-a-year "I'm finally going to": (insert here) quit smoking, quit drinking, stop eating so much, quit dating those go-nowhere types, wasting so much time or money—all-restriction, all-the-time resolutions. What a crazy way to start off a new year of life. And, I don't just mean the insanity of starting your year off with something you have over a 90 percent chance of failing at but also the horrible timing of said failing. For most resolutions, this timing lands somewhere between Blue Monday's onslaught of financial depression and February, the darkest most depressing month of the year, better known as the month where resolutions go to die. Sound like a road to happiness? It seems fairly intense to me.

Now let's take a look at some of the more popular resolutions. First, we'll have to throw out that one that is, in fact, "actually doing my New Year's resolution." (A poll was conducted with the top resolutions for 2020, and among the list, actually sticking to the resolution was the number one goal—but according to a study, only eight percent of people who make New Year's resolutions actually achieve them.)

The typical list goes something like this: exercise more, lose weight and eat healthier, try or learn something new, save money, quit smoking, read more, be more organized, travel, spend more time with loved ones. Then there are the more vague or esoteric ones like live life to the fullest, be a better person, and my all-time favourite, be happy!

Even if we take away the facts that resolutions are usually these huge, lofty aspirations, and that their timing is so depressingly problematic, at their core they are still goals. And as we have already explored, goals do not work for everyone and we are not great at achieving them in general, hence the single-digit success rate mentioned above for goals wrapped up as New Year's resolutions. Add to it the culture of encouragement and forgiveness for everyone bailing on their resolutions early in the year, and you start to see the repetitive pattern of set resolution, fail and repeat again next year, insanity. But hey, it does make us all feel better to fail or quit together, right? Maybe that's the addiction to the timing of these massive once-a-year aspirations.

Now, let's look at the types of things people are trying to accomplish through the eyes of 90-90-ONE. Remember, 90-90-ONE is about failure. Not aspiration—failure. Where in your life are you failing, where are things not working out how you would like, and have you looked into why? Into really why and what clues may be found around this failing?

Are you failing to reach your weight goal because of the lack of the right diet trend? Or is there something else going on, left unresolved, that is tripping you up time and time again when it comes to sticking to a full-on diet? And is there an easier and less daunting way to approach weight loss without having to completely radicalize and limit your food intake? Is there a ONE that can solve this? A ONE that has been there all along, sabotaging your resolve and waiting for you to take notice of it?

The Power of ONEs

Let me digress a bit with an example explaining my thinking on this approach.

When it comes to body weight most health experts will boil things down to calories in and calories out. I know that sounds old school and yes, there is the new, trendy "what type of calories" approach, but for simplicity's sake, let's just keep it at in and out. Now, let's bring in the little-known fact that on average 3,500 unused calories will be stored as one pound of body fat. This knowledge is very powerful when it comes to losing poundage. Look at someone who treats themselves to a chocolate bar as a late afternoon snack once a day. On average, each bar is 250 calories and if that goes unused then every 14 days this individual will add one pound of new fat to their body. Or over two pounds a month. Twenty-six pounds a year. Wow! Hopefully going to get the chocolate bar, opening the wrapper, and chewing use up a few calories and it won't add up that quickly but wow—approximately 26 pounds a year from one simple afternoon indulgence that goes unused.

Okay, so back to the losing weight as a resolution thing. Let's say the goal is losing five to ten pounds and you start it off with a whole bunch of food restrictions and upheaval. If turned instead into a 90-90-ONE, the approach would be to first look at your weight gain or lack of ability to stick with diets and analyze your average day and make note of all the places you think you are failing at losing weight. What are some of the clues to why you fail at keeping the weight off? Is it not keeping activity commitments? Is it giving into cravings for foods you have restricted from your diet? Or not seeing a light at the end of the restriction tunnel and just giving up? What are the clues and what are you 90 percent capable of fixing?

In the above example, if you resolve that the chocolate bar is where you feel you are failing, you decide to make it your ONE. You commit to stop eating it, and not replacing it with something else but also not changing anything else about your diet and calories out. By the way, this is all 90 percent in your control, no need for outside costly shakes arriving in the mail or motivational trainers invading your day. Just you and your commitment to this ONE simple fix, and doing it for 90 days. Assuming the above calorie math is correct, you are down six to seven pounds at the end of this 90-90-ONE! Afterwards, you can regroup, look at something else diet-wise you would like to fix for your next ONE, like going for daily walks, or you can change your 90-90-ONE to another area of your life altogether. As we've seen above, the list for improvement is long. At least it is on January 1st.

I know this sounds simple, but think about it, it's now April 1st, and you have probably stuck to this 90-90-ONE longer than any Goliath resolution in your past. You can feel good about yourself and prepare to tackle something else. I know—being happy! Just kidding. That is always a side benefit of knowing we are in control of our issues. But in all seriousness, the mindset of 90-90-ONE promotes micro-achievements that can shift behaviour, change habits for the better, and help us improve in areas we want, like weight loss, for the long-term, not just the period of restriction that the resolution covers. This is how 90-90-ONE can build lasting success. By looking at what is failing and fixing a ONE we learn about that micro-behaviour and that awareness can be built on for a lifetime.

On a personal note, for me, it was my coffee and ten pounds of belly flab. You see, back in those real estate days of selling new homes in the country, I did a ton of driving. I lived in the urban center but needed to commute every day to the surrounding countryside. At bare minimum, it was a 45-minute drive on the highway, some locations as far away as 90. And every day en route I would stop at the drive-thru donut shop and grab a muffin and a large coffee. I drank my coffee with cream, no sugar. Always had and I loved it that way.

Now, as most of you know, it's hard to control the amount of anything added to our coffee at a drive-thru and our tastes can get quickly addicted to whatever the automatic dispenser has determined the right amount to be. And addicted to this cream in my coffee I was. So much so that I started to up my cream intake in the coffees I brewed at home as well. I could easily go through a litre of coffee cream a week, just from my home brews not including the ones on my way to work.

Fast forward to a few years into this career and I'm starting to put on poundage. Not tons, maybe ten, and usually they would creep on in my less active winter months and some would go away in my more active summer months but all in all the ten around the belly in the winter was making everything a bit tight and uncomfortable. So, diet it was going to be!

While I was looking into all the restrictions I would introduce into my food consumption a colleague asked me if I had ever thought about not putting cream in my coffee, switching it instead to milk. I have to be honest, it never crossed my mind for a second and would never be part of any diet plan I came up with. I loved my coffee that way and hated it whenever someone would make the mistake and give me one with milk. But the seed was planted and I started down the path of discovery around my multiple coffees per day.

Here's what I learned. The difference between half-and-half coffee cream and two-percent milk was approximately an additional 190 calories per cup of half-and-half or coffee cream. Now taking into account my estimate of at least a litre per week that I was consuming, I grabbed my calculator and got the shock of my coffee house life. The difference between choosing one over the other in my coffee was the equivalent of an extra 40,000-plus calories per year. And using the above-mentioned 3,500 equals one pound, that was over 11 pounds per year! In my coffee, just my freakin' coffee!

My awareness had been blown wide open and every coffee I had from that point on was a constant reminder that I needed to stop the cream. And eventually, with the threat of a much more restrictive diet looming in my future, I decided it was time to switch from cream to milk.

Now you noncoffee drinkers, I know what you're thinking: what's the big deal? But you coffee addicts out there, you all know changing up one's coffee is huge. Especially the first one in the morning. For some, it's reprogramming a habit and a taste that you have had your entire adult life. In fact, I know some people that have given up coffee altogether instead of not drinking it the way they love. So yeah, this switch was huge, and for the first couple of days, I just hated it. The taste and joy were stripped from my morning, my drive to work, and my midafternoon rituals, and I wasn't happy about it.

But something amazing happened. About five days in, a colleague unaware of my switch brought me a coffee with cream. With the takeout lid hiding the content's colour, I peeled back the tab and took a sip and was grossed out to taste something so rich and to feel the contents of the fat from the cream on my lips and I realized that I had already lost my taste for rich, creamy coffee. I was now, almost overnight, a milk guy.

That was many years ago, but to this day I still only put milk in my coffee and I will never forget the feeling of slowly having pounds melt off my belly by changing only ONE simple thing. ONE simple micro-behaviour.

Another powerful difference between a once-a-year New Year's resolution and a 90-90-ONE is if you don't achieve it for some reason, you get a do-over. In fact, you get three more do-overs before the year is done, and with each one comes the opportunity to feel good about your progress which, just by this process alone, makes 90-90-ONE perfect for building confidence, efficacy, and agency. And a much better way to accomplish resolution-type goals. You don't fail big and you're done for the year, you get multiple times to succeed, in multiple areas of your life and feel happy. This time, no kidding!

So, using 90-90-ONE instead of massive once-a-year resolutions gives you a way to manage things that are 90 percent within your control, that you believe have been leading to you failing at being the individual you aspire to be. You get to mix and match and do them for a reasonable 90-day period of time four times throughout the year.

If you look at most of the resolutions, almost all of them will have a unique, systemic something that lies in the shadow of that goal that is preventing you personally from becoming someone who reads more or travels more or is a better person. By tackling these issues through the 90-90-ONE technique you build a foundation for more of what you desire by ensuring you will not be tripped up by failure along the way.

Saving ONE at a Time

One mammoth resolution-type goal that I personally have fond memories of turning into a 90-90-ONE was the desire to save more money. This was back before debit and credit tap was so prevalent and it was common to use cash for most small daily transactions.

At the time, I was just starting out in my career and I was struggling with money, like most young, fresh out of college twenty-somethings do. I was living pay cheque to pay cheque and never seemed to have the excess money for special occasions like birthday gifts, first dates, or rainy days. I would try desperately and set lofty goals for saving money that I would never live up to and eventually dive into the meager savings—failing usually within a week or two from starting.

At my wit's end, I decided to try a trick I had heard about. Every time you come home, take your money out of your wallet and collect all the one-dollar bills or coins, depending on where you live, and put them in a place for safekeeping. It was simple. It was 90 percent in my control and I would try it for 90 days.

The surprise for me with this ONE, which turned out to be a bit of a wild card, was that the ten percent that was out of my control was the type of change I would be given back from vendors as I went through my day. I had no control over how many one-dollar bills I would walk into my apartment with at the end of each day. Some days it would be none or one, others a couple, and then there were those crazy days when my change for a coffee from a ten would be mostly ones. I could only control that I put them aside. All of them.

Another surprise benefit of this 90-90-ONE was that I was going through my cash a little bit faster than usual which caused me to visit the ATM a few more times a month than normal and these visits would be a constant reminder of my bank balance and that increased awareness alone created a frugal outlook as I went through my 90 days. I was curbing my everyday spending to compensate for the one-dollar bills that were piling up in the drawer at home.

At the end of the 90 days, I had created a valuable new habit which I have done versions of over the years since, and over

150 dollars in savings I could use for life's little surprises.

Nowadays there are all sorts of apps and cards that round up or add on and can take care of this type of forced savings for you, but as I mentioned at the start of this story, I have a great fondness for this 90-90-ONE because it fundamentally changed how I used money and shifted my mindset to seeing myself as a person who saves. You see, like the diet example before, I don't think a big one-time resolution of saving money in general would have had the same long-term effect. It was turning it into a ONE that help create this micro-behaviour habit that made all the difference to my long-term success.

On an end note, the thing I love the most about breaking resolutions into 90-90-ONEs is the amazing way you will end each year. Imagine, as opposed to cruising through the last three months of the year eating, drinking, and being the worst of yourself; all the while knowing that the new year is looming and with each coming day you are building the case that this year my resolution needs to be huge, meaningful, erase all the accumulating bad behaviours and, most of all, actually stuck to! Remember that one? Oh, and all those successful people we discussed back in Chapter Two. Yeah, that unrealistic comparison is waiting for when the ball drops as well. Instead, by doing 90-90-ONEs throughout the year, you spend the last three months working on something. Fixing something. You end the year with a success and your start to the next year will happen with a pat on the back and another confidence-building, manageable ONE. You get to ring in the new year without having to give birth to some huge task, restriction, or improvement. Why? Because you resolved to look at failures and fix them, little ONEs at a time.

Chapter Eight: Empowering Youth

Them Kids!

Going personal is simpler for people who don't have kids currently in their lives; it's like resolutions as mentioned in the last chapter. But, for people with kids in their lives, "personal" can get very complicated because it's hard to look at your world in isolation. You have the world of your kids to think about as well. In fact, many of your previous goals and resolutions may actually not have been 90 percent yours because they involve your children in some way, shape, or form. It can get really murky and the 90 percent's superpower is removing murky!

Let me reassure you, as someone who has a couple of kids, I know for a fact that 90-90-ONE can and will work as perfectly for them as it does for you personally. Why? Because kids' priorities and behaviour change on a dime all the time and never seem grounded. The flexibility of the 90-90-ONE technique can match their changing world, moods, and priorities. It can also be modified to work with their attention span. Let me explain.

I know there are lots of examples of kids who seem to move

mountains and stay focused in their young lives, whether it's the young child who raises awareness of the injustices of poverty in the world and builds a global charity to make change or the teenager who protests alone outside her parliament building and starts a global climate crisis movement. These teens are incredible, and they have accomplished more in their short time on earth than most adults can do in ten lifetimes, but the truth is, they reflect adult-style achievement back to us and that is what most fascinates us about them. They seem so grown up, mature and powerful for their age. And you know what—they are. Because, for the average kid, life and life's goals and desires change quickly and frequently over time. And when their goals and desires change, at times abruptly, long-term, adult-style tools and techniques for goal setting and achieving can crumble in this upheaval and stress, leaving them feeling less confident.

They need help from tools and techniques and, yes, a mindset that reflects these sudden changes and can adapt to them without causing guilt or shame for not following through with their goals because their process does not fit the paradigm of the adult model of having one's act together. 90-90-ONE does exactly that for them.

Giving them 90 days or even less to focus on something can

fit their attention horizon better. Things like "this year" can seem like forever. Remember how long it took to travel around the sun 365 times when you were a kid? Yes, a life-time. Breaking something down into 90 days can fit and works well for most young people. But, depending on their ONE, you can change the 90 days to reflect your kid's rhythm. For example, you can make it 90 minutes. 90 hours or drop the 90 altogether and make it nine days if you are dealing with a need for flexible timelines. The 90 days is the only thing in 90-90-ONE that can be adjusted downward without causing it to fall apart. That is why it works so well for kids.

Also, and this is where the real magic is, by asking them to look around and make a list of things that they believe are not working and then deciding which they can take 90 per-cent responsibility for fixing is extremely empowering for young people, who so often are told what is wrong and what they need to do by the older generations of parents, guardians, teachers, coaches, and mentors. Their goals are externalized. 90-90-ONE internalizes them. And that is powerful!

I've often been amazed when asking a young person what they think needs to be fixed. How one, their answer is so relevant and obvious; two, it has far-reaching consequences;

and three, it would never be on the list of the top five things an adult would think their teenager should be focused on. Try it. They're very observant creators who are mostly programmed not to speak about their observations. But I digress.

Try it with your kids. Ask them to make a list of all the things that they think are not working well, or causing them to trip up, or are just plain frustrating them. Break the list down into what is within their control to fix, i.e., the 90 percent. The best part about the 90 percent with kids is you filter out and eliminate all the "I need an iPad or my bed needs replacing" kind of distracting stuff. Then give them 90 days, or a suitable version of time as mentioned above, to accomplish it.

Share your ONEs around the dinner table and watch how fast they get done, but also how empowering it is for kids to notice, address and resolve their own issues. It's a mindset that you can develop in your children while they are young that will benefit them for their entire lives. The awareness that "Yes! I can take responsibility to fix the things that I see around me that do not function properly and bring about success and growth." What a wonderful gift to give a child of any age.

It's the Bus's Fault!

Let me share with you a story that reflects the power of 90-90-ONE for kids but also a pattern that so many of us are buried behind, where we can't see the proverbial tree for the forest fire, so to speak.

I have a sister-in-law who experienced an issue with her brain development when she was very young and has been tested to be at the brain development age of around a seven-to eight-year-old. She is a grown woman now. With some supervision, she can live relatively well on her own and she has a job that is subsidized by the government. To get to her job, she has to take a subway and a number of buses, a trip that takes about one hour and ten minutes. Because she is subsidized by the government, her place of employment pays her very little, and they put up with her being late every day.

One day while talking with her, I learned that she was late to work every day by about ten minutes. She knew this was a problem but had no solution.

When I asked why this kept happening she said, "Because the number 52A bus, which takes me there, leaves from the station at 8:55 so it gets me there at 9:10. It's the bus's fault," she claimed defensively. Which had been her excuse to her employer for over five years.

When I asked her to walk me through her journey to work she rattled off, with hyper-focused accuracy, the timing of every single detail: setting the alarm clock for 7:20 a.m., prepping the coffee, taking her shower, walking to the subway, the train's arrival, getting to each bus, when they left and how long they took, finally ending up with her at the infamous 52A bus leaving the stop in the station at 8:55 sharp.

I inquired if there was another 52A bus that left the stop earlier than 8:55 and she proudly stated she knows that they come every 15 minutes but emphasized again how the bus that drops her at that stop does not do so in time to catch the earlier bus so she has to wait for the 8:55 to depart for her work.

I started by walking her backwards all the way to her alarm clock and then asked her, "Have you ever tried to set your alarm clock for 7:10 instead of 7:20 and see what happens?" She responded passionately and shared that she has always gotten up at 7:20 because that was when she was told to get up by her care worker when she first got this job. But after a little discussion, she agreed that setting her alarm clock is within her control and she agreed to try it for a month with the goal of reporting back to me how often she is late during this trial period. Nothing else in her routine changes.

Surprise, surprise, it didn't take a month for her to report back to me that she now arrives every day at 8:55. Five minutes early. And her employer is thrilled to have her there on time. Problem solved! You see, her entire story of how she got to work changed; all the details and times got shifted, and in the end, the 52A bus stopped being the problem.

Now, I know this story is simple and, from an outsider's point of view, obvious, but that is what our own stories are often like. We can be blinded to the real cause of the problem or issue, often stopping at what is closest to the problem but outside of our control as the excuse for why something has to be the way it is or can't be fixed. This is why it is so important to teach 90-90-ONE to young people—because our youth is the exact moment in our lives when this programming of "it's outside of my control" is embedded. It's a time when youth are constantly being told what the rules are and how to be, and like my sister-in-law are just doing things that their guardians told them without looking at whether things are working out in the best way possible. Teaching them to look for things that are not working out and dissecting them for clues to find where they have 90 percent control and responsibility and then diving into that area can shift the entire timeline of a situation or problem and have a powerful positive outcome.

If we give our kids the responsibility and the power to notice, think about, and fix what they are in control of we set them up to live lives of empowerment. It's that simple. By empowering your youth to own their own issues and have a 90-90-ONE mindset they can and will improve their impact in this world and the way the world impacts them.

Chapter Nine: Community—That Committee Thing!

This chapter, I have to say, is a touch on the fringe side of things so I'll keep it short. I know, sadly, that not everyone does community work but the value of this chapter is that most of us have done committee work—and have the emotional scars to prove it! You know that work, that painful, brutal work that involves a group agreeing on how to make something happen? Constantly reminding us all of that famous saying: a camel is a horse designed by a committee.

Where 90-90-ONE has value when it comes to committee work is that it gives us some small, very much required and needed, micro-wins. Let me explain.

The number one thing that bogs down committee work is a word I have learned to be wary of, and that word is "consensus." What does it mean, really? The definition of consensus, when looked up, boils down to "general agreement." We have all agreed to . . . fill in the blank. But by employing the word consensus, committees suggest that there was debate and compromise and some form of coming together, a meeting of minds beyond conflicting views. That work and time and great effort were involved. It's the "12 Angry Men" word to make something seem like this agreement is huge! We had a consensus! It must therefore mean something, right? And yes, it does. But not what we might think. You see, every time you hear the word consensus, look around and try to figure out who just twisted themselves into "compliance," all the while, most likely, harbouring a deep desire to un-consensus the whole damn thing. Because they're most often sitting, jaw clenched, silently fuming right before your eyes. In every consensus there lies, hidden, the passive-aggressive by-product for potential future sabotage. And with that, nothing actually gets done and the circle of committee life loops back around with the soul-depleting need for another meeting and, yet again, another consensus.

Okay, back to community and 90-90-ONE. You see, most community work involves some form of a committee. Whether it's working at your kid's local school on the parent council or trying to make change in the civic world, pretty much all community work involves committees. You know: "We have set up a committee to look into that subject." Where there is committee work there is a debilitating addiction to blowing the 90 percent right out of the water. It is almost impossible for community work to effectively stay within the 90 percent when they are solving any of their challenges. "We will need to involve . . ." or "have we consulted with . . . ?" You name it. This happens, it's understood, because committees are usually trying to accomplish big things that involve outside support. And don't think that corporate boardrooms are that much different. Most boards of directors look and operate a lot like community groups. Members always pride themselves on being the ones who thought to involve so-and-so before we can move ahead with a decision, while never really moving ahead with a constructive active solution.

Now don't get me wrong, there is a ton of value in collaboration and bringing in multiple stakeholders to ensure decisions are inclusive, etc. I get it, but I also know that the number one reason most people will leave any committee-type organization is the fact that nothing gets resolved, done, or generally agreed upon. How often have you left a committee meeting, whether it's with your condo board or your local fundraising charity, and uttered these words: "We talked and talked and agreed to look into this more before we make a decision."

It's a type of unconscious paralysis that permeates anything that involves a groupthink style of conclusion. Anything that makes us feel like we are giving up our independence will cause us to exercise it even more: "I have a suggestion, let's go away and think about this . . . before we come to a consensus." That kind of independence.

Enter the Saboteurs

Here's a little story to illustrate how in large community or committee projects there can be saboteurs hidden in all sorts of places.

I'm sitting in a vegetarian Chinese restaurant waiting for my takeout one cold February day and while I'm waiting, I can't help but overhear a table of four men talking, quite intensely, about a project that was about to begin in the city of Toronto. Something to do with technology, I surmised, because these four were talking IT lingo and were very passionate about the role IT would have in this new project. As I waited and listened, I got more and more intrigued as to what this project might be, it was sounding big and exciting, and as a citizen of Toronto, I was curious as to whether it would affect or benefit me in some interesting way.

The four were so intense in their discussion they didn't even notice their lunches had been served. They just kept on talking passionately through the steam rising up off the hot food. Pausing only to wipe it from their glasses, they continued for about ten more minutes, focused on what I could only assume was a brainstorming session on how to move a mountain.

All of a sudden, one of them proclaimed boldly, "I've got it! If we tell them we will have to shut down, change, elimi-nate . . . something technical that I didn't understand but sounded really important . . . they won't want to risk that and they will kill it." And with that description, the other three agreed, congratulated their colleague for his bril-liance, all concurring that the project would never move for-ward if *that* was at risk, and instantly the intensity of the discussion was released and all four started to dig into their plates of food. They looked happy, relaxed, and content that they had figured out a way to avoid having to do some-thing.

I sat there shocked. Deflated. I had not been eavesdropping on a brainstorming session to move a mountain, I, in fact, had borne witness to the exact opposite. The intense effort to stall a committee from getting something done. And worse, the general acceptance that this type of behaviour was okay. Wow!

The reason I bring this story up is not to say that this is what happens all the time in community or committee work but more to emphasize that a million things can be going on and mucking things up from multiple unseen directions. This can be demoralizing for others, and if you have smart, motivated people leaving because they feel like nothing is getting done then adding a few 90-90-ONES to the team's to-do list to rectify this can't hurt. In fact, they might just be the glue that holds everyone and everything together.

It's in the Planters

One case study that really brings this to life involves my wife and a grassroots community group that was trying to beautify an abandoned local park. A few years back, there was an initiative by our city to fund local park beautification projects. If you had a community group that wanted to put in the effort, the city was handing out grants to make local parks more beautiful. All that was needed was some organizational effort to apply for the grant. My wife was already very active in local grassroots movements, whether saving pools, starting farmers markets, or helping local wannabe council members get elected. You name it, she got herself involved, and along the way she learned what was possible when groups of like-minded folks came together for the betterment of our communities. So she and a few other local, make-things-happen heroes jumped at the opportunity to turn a local hell hole of a lost and forgotten park into a new and improved haven for the growing number of young families that were arriving in our neighbourhood and calling it home.

You see, this park had some real problems. It was part of an old creek system and had steep embankments on three sides, causing ground water to pool and turning the playing field into a soggy mess. The playground equipment was old, broken, and dangerous, and the neglected surrounding area was becoming home to undesirable activities.

So, in went the grant application, and out came the funds to transform the park!

It was a huge win for this small grassroots committee or, as they now branded themselves, "The Friends of Moncur Park." They got a talented local artist to paint a butterfly mural to transform the side of the old run-down utility building, tons of trees were planted around the playing field to help soak up the standing water and the entire playground was rebuilt and modernized. The park had a new life! Families started coming and using it, now feeling safe hanging out with their young children. There were yearly pumpkin walks on Halloween and movies in the park during the summer. The transformation was unbelievable and high fives and big congratulatory pats on the backs were given to everyone involved. The committee members all walked a bit taller knowing that they had, together, transformed an important piece of their neighbourhood.

Fast forward a few years later and the park is doing much better but is still in need of upkeep and additional beautification and The Friends of Moncur Park committee members are still all together and eagerly fighting for more funding and opportunities to continue building out the park's true potential. The focus now is on a bigger, more complex issue. They are now trying to address the old creek system that is responsible for constantly flooding big areas of the playing field. The additional trees have helped but there is still more that can and needs to be done. But hey, they have moved mountains before and they know they can do it again, so they are still in it to win it.

But, as things go in civic politics and budget allocations, money for these types of initiatives was becoming harder to find. Add the complexity of the issue and the multiple players and city departments needing to be involved, like engineering, water management, parks, arborists, etc., and the bureaucratic red tape became mind-numbing, making it easy for the powers that be to say "not now" or "try and apply for this grant next go-around." All the kinds of heartbreaking and demotivating things I outlined before that can happen to community or committee projects and can leave members frustrated were now happening to this team.

Frustrated by constant delays, the Friends decided, in true 90-90-ONE fashion, to do something, anything, that would feel like a win. Like progress. So the team looked around the park and asked, "If beautification is our goal, what do we have 90 percent control over doing that will be done in a short enough period of time—90 days—so we can all feel like we are not wasting our time?" And quickly their eyes and minds landed on three large city-owned concrete planter boxes sitting atop a grassy hill, at street level right beside a couple of benches and a staircase that was one of the entrances to the park. The benches and planters were set back from the street under some trees and the area tended to attract some of the previously mentioned undesirable activity. Because the planters were set back from the street, the city often neglected to include them in their spring plantings of other similar planters throughout the city. These three planters had been left to gather cigarette buts, empty plastic bottles, and worse.

The committee decided to tackle the planter boxes. They took to the internet and with some effort and serendipitous luck they discovered the perfect solution, an affordable tulip bulb program being sponsored by the Netherlands. They took some leftover funds they had remaining from the big makeover a few years back, gathered everyone in the community that had a green thumb or a big heart or both, and got to work.

By the following spring, the planter boxes had been transformed into exploding, full-blown colourful displays of flowering tulips. They'd brought the entrance to the park up to a level of beauty and pride, and they did it easily in the time they would have put aside for heavier, less productive meetings. It gave them all a great feeling of accomplishment. It reinvigorated their desire to continue the harder work and fight with the city and its bureaucracy around the bigger projects but mostly it re-bonded the team and gave them something to feel proud of once again. It was a small thing but because it reflected their bigger purpose it rekindled the driving soul of The Friends of Moncur Park and kept them together and united to see their end goals for the park manifest.

So try it. What are some of the things that are wrong, broken, and need to be addressed that you, as a committee, can solve 90 percent on your own? Pick ONE and now do it over 90 days. Trust me, do 90-90-ONEs for a year and keep track of them, because by the year-end, these ONES might be the only things that actually got accomplished, while the other loftier challenges remain stuck in the consensus mud of committee-focused work. The glue that keeps the committee's valuable members together, engaged, and focused on the bigger, more complicated challenges just might be the ability to point to four ONEs that were accomplished along the way.

Chapter Ten: Conclusion—Start with Simple

A Simple Thing

As we have discovered together, 90-90-ONE is a very simple technique, but like most simple things in life, its significance can slip by and go unnoticed leading to some powerful but simple solutions being overridden and left behind for the more intense, ambitious, and oh so sexy, complicated, mojo-enhancing ego stuff. You know what I mean?

The simple early-morning stretches and walks in nature that give way to the over-priced indoor cycling gear that is connected to WIFI and the world of the internet and all the complications and stresses that go with that. Or the expensive gym membership that creates stress and anxiety around getting there at the precise time when the limited equipment you covet is free for you to step on, get on or under, and use. Only to learn that the latest data on promoting longevity promotes time taken for a good, well-deserved morning stretch and some meditative time walking in nature.

Yes, the power of simple things wins time and time again even when we've forgotten to simply do them.

I digress, but not without purpose. You see, the simple, easy to follow and execute technique that is the 90-90-ONE mindset has some truly amazing superpowers. Like that stretch and walk in nature for longevity, the power of 90-90-ONE is paramount to your future success and contentment along the never-ending growth path of your journey for a successful life. Be it at work or at home. So, let me conclude this time together with a recap of what we have learned along the way with some added backup to remind you that this new mindset will be of great value for you, your team, your kids, and your community. All processed through one simple ONE at a time.

Empowering People

Let me take a second to explain and share my motivation for writing this book. It's best summed up by Tucker Max of Scribe when he asks people pondering whether or not to become authors this question: "If there is just one person, someone who could use your story to map their way out of pain and save themselves, wouldn't that be worth it?" "That pain," in this case, can be subtle but have a huge impact—the feeling of being stuck, or worse, trapped in the failure loop. Living in the polarity of failure and success and not having the tools, language, techniques, or mindset to navigate through the creative iteration journey that is your own personal growth path. You know—that pain!

To help push back on that pain, this book has offered up a bit of a rebel paradigm. You see, the gold is in between, the truth is never obvious. And the paradigm that *failure leaves clues* invites people to be curious, ponder, and start thinking and noticing. And to honour their curiosity in noticing the clues. It's okay to see them. To point them out and help others notice them too. Because they're there.

It's okay to embrace failure. To speak up and ask, what if we looked at this a bit more deeply before we charge forward? Does anyone else see the same things happening over and over again? Are we truly learning everything we can from this growth process? Are these clues telling us there is something we should be listening to? Begging us to sit up and pay attention to? 90-90-ONE encourages you to notice the clues and gives you a technique to address and use them positively in your creative iteration whether personally or as a team.

This book asks you to look at your world and that of your team(s) through a new perspective, a new lens: What is hidden in the shadows, what is not being addressed and what do I or we have the power to resolve? There are life-freeing lessons hidden in our failures but corporate culture, society, and even our own habits encourage us to discard, push through, and shed them as fast as humanly possible. We are shamed into never acknowledging let alone addressing them. This book asks the question: What power is there for all of us in addressing the truth that dares not speak its name—failure?

By addressing the little things, the clues left by failure, we own our greatness. You see, by not allowing the success demons with their outdated patterns of behaviour to make us feel small and act small, but instead having the courage to acknowledge and embrace our failure, integrating our failure as part of our growth, we are owning our wholeness. Not being small but being who we truly are is a definition of integrity. How great is that?

Failures Day

There are a ton of sayings around failure and leaders showing how one manages said "failure" makes a leader great but these sayings all seem to have one thing in common— moving on. They all are written in the past tense. What did they learn from failure? Getting up from it, bouncing back from it. They all seem to describe failure as a place to move away from. We have very little language around working in or with failure. Or, at least we hadn't.

A new consciousness seems to be dawning right now and some great minds are looking into failure and how our lack of understanding it, our inability to embrace it, can be messing up the link from success to mastery, as put by professor and author Sarah Lewis when talking about her book *The Rise: Creativity, the Gift of Failure, and the Search for Mastery*. Lewis explains a couple of very powerful points that I alluded to back in Chapter Four. She explains that "... failure is imperfect.

Once we begin to transform it, it ceases to be that any longer because once we are ready to talk about it, we often call it something else, a learning experience, a trial, a reinvention; it's no longer the static concept of failure." It's a word that doesn't even define the process that we use it to describe. She goes on to say, "The word failure was originally developed to talk about credit—bankruptcy—coming to a dead end. Failure is a term we use but it was never meant for us. You've come to a dead end and cannot go on? We never come to a full end." Our journeys are ongoing, and the word failure muddies the waters and creates all sort of confusion and stigma. So we drop it and "move on" but Lewis goes on to say, "After failure, shame becomes the paralyzing trigger to not being able to start over or anew." And this is because we have driven failure into the shadows, as I explained in Chapter Four. If, subconsciously, it is a dead end, then of course we are done, with nowhere to go from there but into guilt and shame. Again reinforcing the need to create new language around failure. The sooner we own this word failure in a positive way, the sooner we will see all our creative iterations, whether personal, creative, community or business, flourish.

I'm not an academic or an expert, I have no data to dump to prove what I've learned from all my life experiences. So, please, by all means, explore Sarah Lewis's work and others that are diving into failure with academic vigor. Their work is amazing and once you have developed the muscle and desire to tackle your failures with a newfound awareness, you can embrace the 90-90-ONE technique and mindset knowing that its simplistic beauty allows us to build back through all the ill-gotten shame—ONE clue at a time.

By working with and building on the clues from failure, we are proactively continuing on our creative iteration path and actually successfully moving forward and not just "on" from something that was not ours to have in the first place.

The "Should" Mindset

Shame has a sidekick. It's called the "should" mindset, and until you are proactively focusing on, addressing, and owning your failure and working with the clues that it leaves you, you must avoid just mimicking others' success.

Why? By getting sidetracked by what others are doing and falling into and under a "should" mindset, you get stuck looking outside yourself for your power. The infinite loop of shame and "should" pushes your personal power further and further away from you. 90-90-ONE encourages you to look within yourself, at your own actions, behaviours, and environment. To work with your own intimate issues and build your own power from the only place that it ever existed—right inside your very own unique existence.

I talked about fixing things and taking responsibility as being the ego's kryptonite back in Chapter Three. I promised to revisit that statement, and now is the perfect time. After reading through the other chapters, I believe you have a clearer understanding of what I mean by "fixing" and "taking responsibility for."

The concept of 90 percent and being clear on what is a 90 percent ONE asks us to slow down and become aware of our environment, our immediate surroundings, others, and what is and can be impacted by us and us alone or as a team.

This slowing down, this focus on looking at failure, uncovers the clues for where responsibility lies and who can have the most impact on resolving something. To discover who has it under their 90 percent takes an open mind, a growth mindset and is counterintuitive to the blame, shame, and "should" mindset that fuels the ego's desire for protection, quick action, declaration of fault, and constant moving in what seems like a forward direction. Whose fault is it? Next! As we have previously explored.

Taking the time to understand responsibility to its fullest or any process like 90-90-ONE that is built around exploring responsibility may be hard in supercharged ego environments. If you are trying to introduce into your work environment something as simple as 90-90-ONE and you come up against resistance, know that you might just be triggering the spidey sense of a strong ego leader. They might sabotage the process not because they don't think it could be valuable but more because of what the responsibility trigger might expose.

If this happens, don't be discouraged. Just focus on your immediate team and allow others to learn from your own success with 90-90-ONE. Worst-case scenario, you and your team get a lot of things resolved, bond, and learn tons along the way. Best case, it becomes contagious.

Action At Work

Most success-focused, driven leaders and organizations have a tendency to err on the side of activity bias. As defined by WikiBooks: "Activity bias is when an individual, given the choice of taking action or doing nothing, chooses to take action." And, as we have discussed, the ego loves to be in action mode. Activity bias operates on the principle that people want to do something rather than do nothing: "Their choice to take action may result in a less-than-optimal decision being made, but they feel that taking action is what they are supposed to do." This is what the ego lives for. Asking it to slow down enough to truly understand the dynamics of responsibility is crippling for an ego mindset—it's kryptonite! Which brings us full circle back to the premise of this book: 90-90-ONE is a very simple technique, but in its execution alone, it challenges some fundamental Goliaths of paradigms that have to shift. But the magic to do so is in these simple, actionable details.

Activity bias is something that is very easy for the ego mindset to fall into. If something is wrong, then let's jump to the action of fixing it. On the surface, this might seem right, but the bias piece is that jumping into action satisfies the ego that it is doing something and therefore fixing something. But doing is not necessarily fixing or resolving.

The important step that is often missed out with a quick jump to action is the discovery phase. By jumping to fix the issue the responsibility piece is never fully sorted out, understood, and agreed to, and the chosen action will most definitely be misguided. When it fails, the ego, through activity bias once again, will jump to the next action to be taken. And because this behaviour is similar to the chasing-success behaviour of "Next!" the ego can lull itself into a coma while maintaining the bias of being extremely busy doing something. What 90-90-ONE does is slow us down and break things down into individual, unique ONEs where responsibility is fully understood. There is still action, as we mentioned in Chapter Five, as many as 432 actions per year in a company of 100 people. But all can be fully accounted for and traced back to an issue that has been fully addressed. No bias required!

No Longer An Issue

As I've laid out throughout this book, there are benefits to looking into failure and working with the clues and failure seeds that are lying about in your own and your team's lives. There are benefits to helping managers learn more about how their team members tick and how to help them as was discussed in Chapter Six on personalities. There are benefits to empowering people to take control, be in their own power and feel the accomplishment of morale-boosting micro-achievements.

One final benefit I want to discuss is the benefit of releasing and clearing out issues, and its life-changing effect on one's wellbeing and state of mind. By issues I mean all those things that hang around and remind us that we can't relax or enjoy life until we have dealt with *that* issue.

Let me explain this benefit with the help of some experts. First is productivity consultant, David Allen. As he describes in his amazing book *Getting Things Done: The Art of Stress-Free Productivity*, the brain is a great place to come up with ideas but is a horrible place to store them. And by trying to use it as a storage device we cause ourselves all sorts of stress and anxiety. If you haven't read his book, I highly recommend it—it saved my brain.

If we take Allen's advice, the goal is to get things out of our minds, out of our heads so to speak, and into a more productive place for dealing with them. And by doing this we can become more productive while freeing up valuable space in our minds for what it does best—coming up with more ideas.

Now, if we are walking around in an environment that is constantly reminding us of the things that need addressing, i.e., putting issues into our minds, and if this is your work environment with a repeated pattern of not dealing with failure and its clues but you are constantly being reminded that they are there, in the way, and need to be addressed— one can only imagine the potential stress and unconscious anxiety that is building up just by moving through your workday.

This is where 90-90-ONE dovetails nicely with Allen's thinking and recommendations because instead of address- ing issues that you already have in your head, hindering your productivity, 90-90-ONE gives you a technique to ad- dress the issues that lie in the shadows, as yet unnoticed, waiting to trip you up, create failure and become issues that end up in your mind, potentially hindering your productiv- ity.

The technique is about getting things out of the shadows, out of the shame and blame toxic pool, and definitely out of the mind and dealt with. Helping our productivity with the added benefit of clearing our minds to be more creative. Beautiful!

The next expert I want to lean on around this benefit of freeing our minds comes filtered through Tim Ferriss. In an episode of *The Tim Ferriss Show Podcast,* he explained his take on some concepts in Bruce Tift's book *Already Free: Buddhism Meets Psychotherapy on the Path of Liberation.* Ferriss explains the empowering effect of looking at issues that may have bothered you your entire life. By spending time focusing on them you can come to the realization that they are no longer issues that require fixing or adjustment. You can live with them. As Ferriss went on to make clear: "You can come to terms with the fact that you may feel a certain way your entire life and there is no need to fix anything, and how this has a huge psychological release. The release that will change how you fixate and posture on something and you will see a change/shift for the better." And this same shift from problem to choice is a by-product benefit of the 90-90-ONE technique and mindset.

Let me clarify. By going through a list of issues and determining what are yours to resolve—i.e., 90 percent yours—and what ONEs you want to focus 90 days of your life on resolving you can free yourself from the psychological weight of issues that are not yours to resolve. Even better, you may conclude that something is actually okay as it is, not worth the 90 days, and you release yourself from seeing it as a problem. You will start to see it as a choice you have made to allow it to exist. No longer an issue or an obstacle, it will be a choice. Freeing, right? And by just going through the process of 90-90-ONE you can release a number of these issues all the while taking care of a ONE that you have determined *is* worth the 90 days and that will shift the pattern of your own personal or team growth.

Empowerment

The goal is and always should be to empower people, and by championing 90-90-ONE you are empowering your team, kids, community, and yourself to become successful and to take on goals and set them properly because you have helped figure out how to free yourself and them from all the stumbling blocks that hold us back.

Our lack of a healthy relationship with failure, lack of language around it, and our disconnect between failure and the role it plays in the growth process that is seen as success keep us stuck. But until we as a culture and society learn about failure and develop as a group our new relationship with it you as an individual can practice your own simple approach to better accomplishing growth. Healthy meaningful growth, by using 90-90-ONE. Think of it as a way to reprogram the brain. Getting you to see the things around you that are interfering with your and your team's advancement and take ownership and control of them.

To recap: Why is it so crucial to have a technique, tool, mindset to work with failure? Because as a culture and society, especially in business, we have shoved failure and its clues into the shadows and by doing so we have gotten stuck in a repeat pattern of what is called "unconscious patterning" around our failures. When it comes to the unconscious, experts conclude, we will constantly put ourselves in situations to create the same problems so that, eventually, (or so we hope) we will learn the required lesson. Until that time, the issues resurface over and over again. They change their look and their feel and how they present but underneath is a fundamental problem that we are not addressing. By looking at our failures and looking for clues we are making them conscious and breaking this pattern.

Being Present

90-90-ONE also helps you stay present in the now. The power of noticing and then addressing something no matter how small empowers you by being in control. Fleeing the issue and chasing success fools you into believing that you are doing something and feeds the ego but not the process of your own iteration and growth and definitely does nothing to remove the issue at hand from resurfacing again and again.

When it comes to goals, this book has asked you to take another look at them. Focus on the system and process of achieving and not achieving them and not just on the end goal. By refining and building the process of establishing goals the outcome will appear and it will be different than before.

Refining is the art of looking deeply at your failures. Why are they happening, what might be the clues? Then cleaning up the little things you find along the way, not just manically jumping from goal to goal in a success-driven fever dream. It's taking the time to look inward. At yourself, your goal, your failure, and what there is to learn from it all.

I talked about the work being done by James Clear and outlined in his book *Atomic Habits: An Easy & Proven Way to Build Good Habits & Break Bad Ones* and how he explains that the winning and losing team both had the same goal at the start of the game, which was to win. But if all professional sports teams did was jump from finite goal to finite goal, never refining those goals, never looking for clues to why they are *not* winning then the process of shifting the goal from winning to losing well—like resting starters to avoid unnecessary injury near the end of a losing game or endlessly reviewing game footage to break down how each individual player can set personal goals to improve—would never happen. But we know they refine and review and focus on that failure and come the next game the overall goal of winning is now being backed up by the removal or improvement of all the little clues that failure left for them to refine in the process of becoming a winning team.

This book was written for anyone who is struggling to make something happen. Make something turn out well. It's for that one person who has dedicated their life to make something happen for themselves and/or others. 90-90-ONE is a mindset for the one person that knows that it is okay to say, "I don't know yet," because that is where true self-esteem and confidence lies.

The confidence to know that failure has something to teach and is a real gift long before it becomes socially acceptable and justified through success's rearview mirror. Mining failure is where the courageous go and if 90-90-ONE can help in that journey, I am grateful to have been given the opportunity to share it with you, that "one person."

And to quote the great Maya Angelou "... people will forget what you said, people will forget what you did, but people will never forget how you made them feel." So with that in mind, I hope 90-90-ONE makes you feel empowered.

Acknowledgments

Susana Molinolo, Sebastian and Savannah Gahan, Helga Marizel, Michelle Piller, Sue Cooperstock, Joyce Curry, Martin Millican, Vince Font, Alan C. Logan, Murray Reiss, Rachael Muir, Marylin Lacroix, Lee Parpart, Tim Farriss, Brené Brown, Sarah Lewis, and last but never least, The Dream World.

About the Author

Brian Gahan's career has taken him through a multitude of industries: restaurant, construction, head-hunting, real estate, digital & dot com, advertising, and podcasting. He has travelled to more than 30 counties, worked as a photographer in some, and constantly reinvented himself along the way. As a student of life and learning, he developed a fascination with the concept of failure and started to explore the chasm between learning from it and running from it, and why most businesses are stuck in the latter. And then he wrote a book about it.